MEDIEVAL MODAL SYSTEMS

This book explores noteworthy approaches to modal syllogistic adopted by medieval logicians including Abélard, Albert the Great, Avicenna, Averröes, Jean Buridan, Richard Campsall, Robert Kilwardby, and William of Ockham. The book situates these approaches in relation to Aristotle's discussion in the Prior and Posterior Analytics, and other parts of the Organon, but also in relation to the thoughts of Alexander of Aphrodisias and Boethius on the one hand, and to modern interpretations of the modal syllogistic on the other. Problems explored include: Aristotle's doctrine of modal conversion, the pure and mixed necessity-moods, modal ecthesis, the pure and mixed contingency-moods, and Aristotle's use of counter-examples. Medieval logicians brought various concepts to bear on these problems, including the distinction between *per se* and *per accidens* terms, the notion of essential predication, the distinction between *ut nunc* and *simpliciter* propositions, the distinction between *de dicto* and *de re* modals, and the notion of ampliation. All these are examined in this book.

ASHGATE STUDIES IN MEDIEVAL PHILOSOPHY

Series Editors

The study of medieval philosophy is flourishing as never before. Historically precise and philosophically informed research is opening up this large but still relatively unknown part of philosophy's past, revealing – in many cases for the first time – the nature of medieval thinkers' arguments and the significance of their philosophical achievements. Ashgate Studies in Medieval Philosophy presents some of the best of this new work, both from established figures and younger scholars. Chronologically, the series stretches from *c*.600 to *c*.1500 and forward to the scholastic philosophersof sixteenth and early seventeenth-century Spain and Portugal. The series encompasses both the Western Latin tradition, and the Byzantine, Jewish and Islamic traditions. Authors all share a commitment both to historical accuracy and to careful analysis of arguments of a kind which makes them comprehensible to modern readers, especially those with philosophical interests.

Other titles in the series:

Theology at Paris, 1316–1345
Peter Auriol and the Problem of Divine
Foreknowledge and Future Contingents
Chris Schabel
ISBN 0 7546 0204 4

Medieval Modal Systems

Problems and Concepts

PAUL THOM
Southern Cross University, Australia

Routledge
Taylor & Francis Group
LONDON AND NEW YORK

First published 2003 by Ashgate Publishing

Published 2017 by Routledge
2 Park Square, Milton Park, Abingdon, Oxon OX14 4RN
52 Vanderbilt Avenue, New York, NY 10017

Routledge is an imprint of the Taylor & Francis Group, an informa business

British Library Cataloguing in Publication Data
Thom, Paul
Medieval modal systems : problems and concepts. – (Ashgate studies in medieval philosophy)
1. Modality (Logic) 2. Logic, Medieval
I. Title
160

Library of Congress Cataloging-in-Publication Data
Thom, Paul.
Medieval modal systems : problems and concepts / Paul Thom.
p. cm. – (Ashgate studies in medieval philosophy)
Includes bibliographical references and index.
1. Modality (Logic)–History–To 1500. 2. Logic, Medieval.
I. Title. II. Series.
BC35.M63T48 2003
160'.9'02–dc21 2003043681

ISBN 13: 978-0-7546-0833-2 (hbk)

Contents

List of Figures

List of Tables

Preface

The major medieval systems of modal logic were developed during a period flanked by Pierre Abélard in the twelfth century and Jean Buridan in the fourteenth. Little of great significance pre-dates Abélard or post-dates Buridan; and these two Frenchmen are major figures indeed. In between their work lies that of the well-known Avicenna, Averröes and Ockham, as well as that of the comparatively unknown Robert Kilwardby in the thirteenth century and Richard Campsall in the early fourteenth. The work of these seven thinkers in the field of modal syllogistic will be my focus in this book.

These thinkers were motivated by two forces. First of all, there are the interpretive puzzles posed by Aristotle's modal logic as expounded in the *Prior Analytics* and *De Interpretatione*. These two texts appear not always to be in accord with one another, and the modal syllogistic as presented in the *Prior Analytics* not only lacks a semantic foundation but appears to be internally inconsistent. All this is grist to the mill of the interpreting mind. Then there is the inherent fascination of modal logic as a field of theoretical enquiry. An uneasy relationship exists between the desire to interpret Aristotle and the desire to theorize modality. The latter, of course, is a philosophical desire, autonomous but also intimately connected with broader metaphysical and theological ways of thinking. The former force is, at first sight, a philological rather than a philosophical one. But in reality matters are not so simple, since the philosophical interpretation of any text – let alone one that increasingly carried such authority as that of Aristotle – is always partially governed by philosophical rather than philological imperatives.

There are two key theoretical questions stirring the minds of medieval modal logicians. The first question concerns the doctrine that I shall call actualism. According to this doctrine, modal propositions are about the actual things that ordinary non-modal propositions are about. Contrasted with actualism is the view that modal propositions are about what *possibly* falls under the subject-terms of the corresponding non-modal propositions. This doctrine I will call ampliationism. The leading actualists in the medieval period are Abélard, Ockham, and to a certain extent Campsall; the ampliationists are Avicenna and Buridan. A second question concerns the extent to which essentialist notions are assumed in the modal theories of our seven thinkers. All seven make use of the notion of an essential property. In addition, Averröes, Kilwardby and Campsall make use of the notion of a kind, i.e. a class whose members necessarily share their essential properties.

Acknowledgements

Some parts of this book have seen earlier incarnations. Chapter One draws on the results of my book *The Logic of Essentialism: an interpretation of Aristotle's modal syllogistic*. Earlier versions of Chapter Three were delivered at the Philosophy Department of the University of Queensland, at the conference 'Pierre Abélard, à l'aube des universités', held at the University of Nantes and at a mini-conference on the History of Logic held at the University of Helsinki (2001). Earlier versions of Chapter Four were delivered at a History of Logic Conference held at the University of Cincinnati and at the Mulla Sadra Conference in Tehran (1999). An earlier version of Chapter Six was delivered at the Philosophy Department of Uppsala University (2001). An early version of Chapters Eight and Nine was delivered at the Ancient Philosophy Conference held at the Australian National University (1995). I would like to record my fond recollection of those learned discussions, and my gratitude to all who participated in them, especially Fred Johnson, Simo Knuuttila, Henrik Lagerlund, Constant Mews and Calvin Normore.

For her constant support and infectious good humour I thank Cassandra.

List of Abbreviations

L	Necessity
M	Possibility
Q	Contingency
X	Actuality
→	'is included in'
←	'includes'
|	'excludes'
⌣	'overlaps'
*	'necessarily' (governing a term)
†	'possibly' (governing a term)
‡	'contingently' (governing a term)
□	'it is necessary that'
◊	'it is possible that'
∇	'it is contingent that'
~	Negation

Chapter 1

Aristotle's Modal Logic

To begin at the beginning, we need to understand the essentials (so to speak) of Aristotle's modal syllogistic, that being the foundation on which the whole of medieval modal logic is built. But no foundation is ultimate.

The Platonic background

Many ideas in Aristotle have their antecedents in Plato. In the case of the modal logic, we can find several such antecedents in Plato's *Phaedo*, where Socrates is portrayed as traversing a three-stage intellectual journey. At first he held that a thing may be beautiful by virtue of the gold in it. Then he thought that a thing could be beautiful only by virtue of the beauty in it, and analogously that a thing could be hot only by reason of the heat in it. Finally, he recognized – in the case of heat – that something can be hot because of the fire in it (because fire is by its nature hot). A metaphysical logician reading this, may well have been set thinking about what is in effect a modal syllogism, concluding that heat inheres in a subject, on the ground that heat necessarily inheres in fire, and fire inheres in the subject.

Towards the end of the dialogue, Socrates advances his final argument in favour of the soul's immortality, and this argument (so it happens) depends on a number of notions that will later resonate in modal syllogistic. Key among those notions are the following three.

1. The forms are supposed to exist and other things take their names from them;[1]
2. Snow cannot be hot, while remaining snow, and similarly fire cannot be cold while remaining fire;[2]
3. While the opposition between heat and cold is, we could say, a primary or nomic one, there is a secondary opposition between fire and snow such that the snow retreats or else is annihilated by the advance of heat, and in general:

---

[1] Plato (1993) 102b.
[2] Plato (1993) 103d.

> ...it is not only the opposite that doesn't admit its opposite; there is also that which brings up an opposite into whatever it enters itself; and that thing, the very thing that brings it up, never admits the quality opposed to the one that's brought up.[3]

Each of these notions undergoes significant development at Aristotle's hands, and each one of them will play a role in his modal syllogistic.

Aristotelian logic and metaphysics

The modal logicians of the Middle Ages saw their enterprize as linked with a motley assemblage of Aristotelian texts, drawn from the *Categories*, *De Interpretatione*, *Prior* and *Posterior Analytics*, the *Topics* and the *Sophistical Refutations*. The fundamental ideas in those texts are the idea of a term, a being, a categorical proposition and essential predication.

Terms

A term is a noun-phrase – any expression that could be subject or predicate in a categorical proposition. A term may or may not stand for a being, and accordingly even an expression like 'non-being', which explicitly excludes beings, is a term.[4] Nonetheless, terms standing for beings constitute an important sub-class among terms. Beings, according to the *Categories*, fall into ten classes or categories. The first category comprizes the substances, and substance-terms notably include the names of natural kinds – terms like 'man', 'horse'. Other categories are the qualities (including colours such as whiteness),[5] and the relatives (including positions such as standing and sitting).[6]

Some terms do not directly designate anything that is in one of the categories. Some terms, for instance, are *denominative*, like 'literate' and 'brave' which, Aristotle states, are expressions deriving from 'literacy' and 'bravery' but with a different ending.[7] Literacy (as a form of knowledge) and bravery (as a virtue) are beings in the category of quality;[8] but the literate and the brave as such are not, even though the people who are

---

[3] Plato (1993) 103d-105a.
[4] Aristotle (1989) A38, 49a23-24.
[5] Aristotle (1963) ch.8, 9a28ff.
[6] Aristotle (1963) ch.7, 6b11ff.
[7] Aristotle (1963) ch.1, 1a12ff.
[8] Aristotle (1963) ch.8, 8b29-35.

literate or brave are beings in the category of Substance. Denominative terms include 'white', 'standing' and 'sitting' (understood as adjectives), though the terms for the qualities or relatives from which these adjectives derive do stand for beings in the categories of Quality or Relative. Denominative terms, especially Aristotle's paradigm 'literate', will figure in the arguments of the medieval modal logicians.

At first sight, Aristotle's definition of denominative terms in the *Categories* seems hospitable to Platonism. It seems almost like a quote from Plato's description of the way in which the name of a form gets applied to particulars that participate in the form. But there is a subversive element in Aristotle's thinking here. Denominative terms may be semantically secondary to the abstract terms from which they derive, but according to Aristotle the primary realities are not the qualities and relatives denoted by those abstract terms but the substances underlying the denominative terms. On this point he is at odds with Plato who took the abstract terms from which denominative terms derive grammatically to stand for the primary realities.

Another group – compound terms – combine a number of terms, some of which may be denominative, for instance 'musical Miccalus', 'intelligible Aristomenes'.[9] Aristotle holds that in some cases the components in a compound term form a unity only accidentally. He gives the example of the terms 'white' and 'musical':

> Of things predicated, and things they get predicated of, those which are said accidentally, either of the same thing or of one another, will not be one. For example, a man is white and musical, but 'white' and 'musical' are not one, because they are both accidental to the same thing.[10]

The crucial thing here is whether the component terms are related to one another *per se* or *per accidens*.

Aristotle applies the *per se/per accidens* distinction to terms as well as to relations between terms. The distinction between *per se* and *per accidens* terms is explained in the *Posterior Analytics* as the third of four senses of '*per se*' [*kath' hauto*]. Something is *per se* in this sense when it:

> ... is not said of some other underlying subject – as what is walking is something different walking (and white), while a reality, and whatever signifies some this, is just what it is without being something else.[11]

---

[9] Aristotle (1989) A33, 47b21-22, 29-31.
[10] Aristotle (1963) 21a7-11.
[11] Aristotle (1975) A4, 73b6-8.

Putting this together with the doctrine of the *Categories*, it seems that a *per se* term is a general term in the category of substance. However, there is also a more general sense in Aristotle according to which a general term in any category is *per se*.

One term may be related to another in a variety of interesting ways, four of which are theorized in Aristotle's doctrine of the *predicables* – definition, genus (and differentia), peculiarity and accident. These constitute four types of relation between a predicate and its subject (assumed to be a *per se* term). The predicate may be either essential or non-essential (*per se* or *per accidens*) to that subject, and may independently be either convertible or non-convertible with it. This generates four possibilities according to whether the predicate is (1) convertible and essential (a definition), (2) non-convertible and essential (a genus or differentia), (3) convertible and non-essential (a peculiarity), or (4) non-convertible and non-essential (an accident).[12]

According to this definition, an accident is what is predicated non-essentially and non-convertibly of the subject; according to a second definition, an accident is what can belong and can not belong to the subject.[13] That the two definitions are not equivalent seems to be shown by the relation of the predicate 'white' to the subject 'swan': this predicate is not convertible with 'swan' and is not essential to 'swan', and yet it is not true that swans can be other than white (or at least Aristotle and his contemporaries thought so). The case is similar with the whiteness of snow (and indeed – to recall the *Phaedo* – with the coldness of snow or the hotness of fire). Which definition, then, is to be preferred? If we adopt the second definition, and if the classification of the four predicables is to be exhaustive, then there can be no such predicate as a necessary accident, because accidents are all *contingent* predicates of their subjects. On the contrary view, we can maintain the first definition of 'accident' whereby the essential is a proper subclass of the necessary, some accidents are contingent and some are necessary, and there is nothing to prevent one and the same predicate being a contingent predicate of some subjects and a necessary predicate of others.

Propositions

A categorical proposition is, according to Aristotle's definition, 'an expression which affirms or denies something of something'[14] – where that

---

[12] Aristotle (1928) A8, 103b6-19.

[13] Aristotle (1928) A5, 102b4ff.

[14] Aristotle (1989) A1, 42a16-17.

which is affirmed or denied is the predicate and that of which it is affirmed or denied is the subject. The proposition is either affirmative or negative, and is either universal or particular, thus generating four propositional forms. In the Middle Ages the four forms were represented by the four vowels *a* (universal affirmative), *e* (universal negative), *i* (particular affirmative), and *o* (particular negative). Among the categorical propositions that Aristotle asserts or supposes to be true in the *Prior Analytics* are 'Every animal is moving',[15] 'Every man is an animal',[16] 'Every animal is waking',[17] 'No animal is moving',[18] 'Nothing white is an animal',[19] 'No horse is good',[20] 'No horse is waking',[21] 'Nothing white is waking',[22] 'Something white is not an animal',[23] and 'Something white is not waking'.[24] In one passage Aristotle identifies a special class of assertorics, ones whose truth is not limited to any time but are true *haplōs* [*simpliciter*].[25]

Of special interest to us are modal propositions. These are categorical propositions to the effect that something belongs necessarily, or possibly, or contingently, to a subject. Contingency is understood as two-way possibility. We have to note that Aristotle distinguishes two ways of understanding contingency-propositions, either as '*a* is contingent for all that to which *b* belongs' or '*a* is contingent for all that for which *b* is contingent'.[26] Following medieval usage, I will refer to the first of these senses as unampliated and the second as ampliated. In the *Prior Analytics* Aristotle's examples of true necessity-propositions include 'Every man is necessarily an animal',[27] 'Every horse is necessarily an animal',[28] 'Something white is necessarily an animal',[29] 'Some man is necessarily an

15 Aristotle (1989) A9, 30a23-25, 28-33; 30b2-6; A11, 32a4-5.
16 Aristotle (1989) A10, 31a15-17.
17 Aristotle (1989) A11, 31b27-33.
18 *Ibid.*
19 Aristotle (1989) A10, 30b7-9, 31-40.
20 Aristotle (1989) A11, 31b4-8.
21 Aristotle (1989) A11, 31b8-10.
22 Aristotle (1989) A11, 31b40-32a4.
23 Aristotle (1989) A10, 31a10-15.
24 Aristotle (1989) A11, 31b40-32a4.
25 Aristotle (1989) A15, 34b6-8.
26 Aristotle (1989) A13, 32b25ff.
27 Aristotle (1989) A9, 30a23-25, 28-33; A10, 30b7-9, 31-40, 31a10-15.
28 Aristotle (1989) A11, 31b4-8.
29 Aristotle (1989) A9, 30b2-6; A11, 31b40-32a4.

animal',[30] 'Some animal is necessarily biped',[31] 'Something white is necessarily not an animal',[32] and 'Some animal is necessarily not biped'.[33] His examples of true contingency-propositions include 'Every man may or may not be white',[34] 'Every animal may or may not be white',[35] 'Something white may or may not be an animal',[36] 'Every horse may or may not be moving',[37] 'Every man may or may not be moving',[38] 'Every cloak may or may not be white',[39] 'Every animal may or may not be healthy',[40] 'Every horse may or may not be healthy',[41] 'Every man may or may not be healthy',[42] and 'Every man may or may not be sleeping'.[43]

Table 1.1 Modal equipollence

I	II
'possible to be'	'not possible to be'
'not impossible to be'	'impossible to be'
'not necessary not to be'	'necessary not to be'
III	IV
'possible not to be'	'not possible not to be'
'not impossible not to be'	'impossible not to be'
'not necessary not to be'	'necessary to be'

In the *De Interpretatione* Aristotle recognizes that modal expressions are equipollent to one another as shown in Table 1.1.[44]

---

[30] Aristotle (1989) A10, 30b7-9.
[31] Aristotle (1989) A11, 31b31-33.
[32] Aristotle (1989) A10, 31a15-17.
[33] Aristotle (1989) A11, 32a4-5.
[34] Aristotle (1989) A17, 37a2-3.
[35] Aristotle (1989) A15, 35a20-24.
[36] Aristotle (1989) A15, 35b11-19.
[37] Aristotle (1989) A15, 34b10-17.
[38] Aristotle (1989) A15, 34b10-17.
[39] Aristotle (1989) A15, 35b11-19.
[40] Aristotle (1989) A18, 37b35-38.
[41] Aristotle (1989) A18, 37b35-38.
[42] Aristotle (1989) A18, 37b40-38a2.
[43] Aristotle (1989) A22, 40a35-38.
[44] The presentation follows that of Ackrill in Aristotle (1963) p.62. Ackrill *ad* 22a14 points out that this is Aristotle's second attempt at diagramming these equipollences, and involves a correction of the earlier attempt.

Aristotle also includes 'contingent to be' and its internal and external negations in the table, and clearly intends 'contingent' to mean the same as 'possible', though in the *Prior Analytics* he differentiates contingency as two-sided possibility.[45] In that text he distinguishes two types of contingency – that which is not necessary but happens most often and naturally (such as for humans to go grey), and that which happens or fails to happen by chance, such as for humans to walk.[46]

The link between the two terms of a proposition may be either *per se* or *per accidens*. The *Posterior Analytics* explains that one term is predicated *per se* of another when it enters into the other's definition or *vice versa*.[47]

Sometimes it is unclear what proposition is to be understood as being expressed by a given sentence, because the sentence is ambiguous; and one type of ambiguity is especially relevant to modal logic – the ambiguity of scope that Aristotle called the fallacy of composition. In the *Sophistical Refutations* he discusses the sentences 'It is possible for the sitting to walk' and 'It is possible for those not writing to write'. Each sentence contains a denominative word (which I have translated as 'the sitting' and 'those not writing'), followed by an infinitive with an opposed sense. Aristotle points out that the sentences are ambiguous. On the one hand the modal expression may be inserted between the opposed words ('For the sitting it is possible to walk') *dividing* them from one another; and then the sense will be that the ability to walk or to write is attributed to someone who is not sitting or not writing. On the other hand, the opposed words may be *compounded*; and then the sense will be that what is possible is that those who are sitting are walking, or that those who are not writing are writing.[48] This Aristotelian analysis is fundamental to much of medieval modal logic.

Another important ambiguity is that unqualified [*haplōs*] statements of necessity differ from necessity-statements like 'Everything is necessarily is, so long as it is'.[49]

The relations between universal propositions and particulars of the opposite quality is one of *contradiction*; that between two universals of opposite quality is *contrariety*. Two particulars of opposite quality are called *sub*-contraries. Universal propositions imply the corresponding particulars.[50] The *De Interpretatione* contains Aristotle's explications of the

---

[45] Aristotle (1989) A13, 32a18-20.
[46] Aristotle (1989) A13, 32a4-13.
[47] Aristotle (1975) A4, 73b5ff.
[48] Aristotle (1928) 166a22-30.
[49] Aristotle (1963) ch.9, 19a23-28.
[50] Aristotle (1989) A4, 26b14-16.

important relations of contrariety and contradiction.[51] The universal affirmative and particular negative are contradictories, as are the universal negative and particular affirmative. The two universals are contraries. The fundamental logical relations (implication and contradiction) among categorical propositions that share a subject and predicate are shown in Figure 1.1. Conceptually, the relations of contrariety and subcontrariety are derivative on these. A contrary of *p* is a proposition implying *p*'s contradictory, and a subcontrary of *p* is a proposition implied by *p*'s contradictory.

From Table 1.1 we can infer the logical relation that holds among pairs of singular modal propositions. For instance we can see that 'It is necessary that *x* is *a*' is contradictory to 'It is possible that *x* is not *a*'. What we cannot infer is the logical relations between two *quantified* modal propositions, such as 'It is necessary that every *b* is *a*' and 'It is possible that no *b* is *a*'. There is indeed evidence that Aristotle clearly understood what these logical relations are in individual cases;[52] but what is lacking in his discussion is a synoptic view of them, such as he presented for singular propositions. For such a synoptic view we must wait for Abélard.

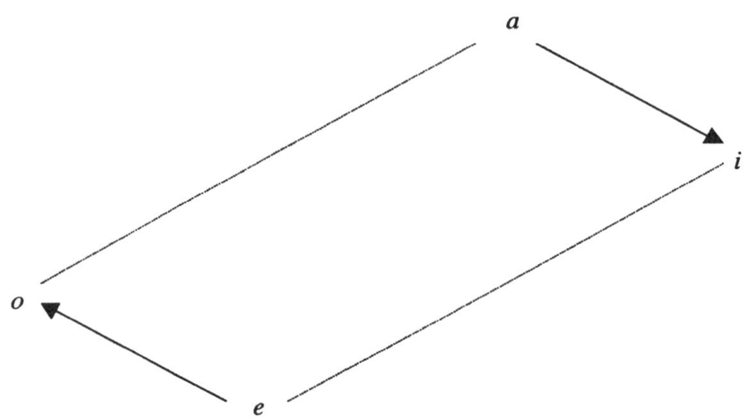

Figure 1.1 Non-modal square of opposition

The *Prior Analytics* outlines the ways in which the subject and predicate of a categorical proposition can be interchanged while preserving truth.[53]

[51] Aristotle (1963) 17b3-6, 16-20.
[52] See Thom (1996) §2.1.
[53] Aristotle (1989) A3.

This is the doctrine of conversion. Universal negative propositions ('No *b* is *a*') are convertible ('Therefore no *a* is *b*'), as are particular affirmatives ('Some *b* is *a* therefore some *a* is *b*'). Consequently universal affirmatives ('Every *b* is *a*') are partially convertible ('Therefore some *a* is *b*').

The conversion of necessity-propositions proceeds just as for non-modal propositions. The universal and particular affirmative both convert to a particular affirmative, and the universal negative to a universal negative.[54] Contingency-propositions behave differently. Affirmatives and negatives are inter-changeable; and while universal or particular affirmatives convert to a particular affirmative, the universal negative does not convert.[55] Bearing in mind that negatives are equivalent to affirmatives, a counter-example to the convertibility of the universal affirmative is also a counter-example to the convertibility of the universal negative. Aristotle supplies such an example, taking it as true that every man may or may not be white (i.e. pale), but false that everything that is pale may or may not be a man.[56]

Syllogisms

If two categorical propositions sharing a (middle) term are combined with each other, they fall into one of three *figures* depending on whether the middle term is (1) subject in one proposition and predicate in the other, or (2) predicate in both, or (3) subject in both. In each figure, there are some combinations of premises that imply a conclusion containing the terms in the premises other than the middle. These are the *syllogisms*. Leaving aside derivative cases where a premise is strengthened or a conclusion weakened, the fundamental syllogisms[57] are listed in Table 1.2.

One syllogism can be equivalent to another, in the sense that it is valid if and only if the other is valid. Aristotle uses two techniques for demonstrating these equivalences – direct and indirect reduction. Both types of reduction involve the transformation of one syllogism into another by means of applying a rule that preserves validity. In the case of direct reduction the transformation involves replacing a premise or conclusion by its equivalent converse. In the case of indirect reduction the transformation is more complex: one premise is left constant, the contradictory of the other

[54] Aristotle (1989) A3, 25a29-33.
[55] Aristotle (1989) A13, 32a29-b1; A3, 25a40-b3; 25b16-17.
[56] Aristotle (1989) A17, 37a4-8.
[57] For this expression see Thom (1981) §29.

premise becomes the conclusion, and the contradictory of the conclusion becomes a premise.[58]

Table 1.2 Mnemonic names for syllogisms

Figure	*Mood*	*Name*
1	aaa	Barbara
	eae	Celarent
	aii	Darii
	eio	Ferio
2	eae	Cesare
	aea	Camestres
	eio	Festino
	aoa	Baroco
3	aii	Datisi
	iai	Disamis
	eio	Ferison
	oao	Bocardo

Table 1.3 Aristotle's system: L/M combinations

Figure	*Combination*
1	LLL,[59] [LMM],[60] [MLM][61]
2	LLL, [LMM],[62] [MLM][63]
3	LLL, [LMM],[64] [MLM][65]

---

[58] Aristotle seems to take the procedure of direct reduction as obvious. He states the principle underlying indirect reduction at Aristotle (1989) B8, 51b1-5.

[59] All the LLL moods are asserted at Aristotle (1989) A8, 29b36-30a2.

[60] These are equivalent by indirect reduction to the second-figure LLL moods. Aristotle does not assert these moods, however at Aristotle (1989) A16, 35b38-36a2, 15-17, 39-b2 he does assert the first-figure LQM moods.

[61] These are equivalent by indirect reduction to the third-figure LLL moods. Aristotle does not assert these moods, however at Aristotle (1989) A16, 36a39-b2 he does assert Ferio QLM.

[62] These are equivalent by indirect reduction to the first-figure LLL moods. The LMM moods are not asserted by Aristotle, but at Aristotle (1989) A19, 38a16-18, b25-27 he does assert Cesare and Festino LQM.

[63] These are equivalent by indirect reduction to the third-figure LLL moods. They are not asserted by Aristotle, however at Aristotle (1989) A19, 38a21-26 he does assert Camestres QLM.

For purposes of abbreviation, I shall use the letters 'X', 'L', 'M' and 'Q' respectively to indicate assertoric propositions, necessity-propositions, possibility- and contingency-propositions. Aristotle's modal syllogisms fall into eight layers, distinguished by the combinations of modalities that occur in them. The first layer is characterized by the presence of necessity- and/or possibility-propositions. Aristotle accepts the syllogisms displayed in Table 1.3. He takes the first-figure LLL moods to be perfect. The LLL moods in the other figures (apart from Baroco and Bocardo) he reduces to the first figure by means of necessity-conversion. Baroco LLL is proved by *ecthesis*. If some *c* cannot be *a*, then we may take some term '*d*' such that every *d* must be *c* and no *d* can be *a*; if then we also assume that every *b* must be *a*, we have the premises of a Camestres LLL: 'Every *b* must be *a*, and no *d* can be *a*'. And from these premises we can infer that no *d* can be *b*. But since every *c* must be *d*, we can further infer that some *c* cannot be *b*. A similar proof is available for Bocardo LLL. The remaining moods in Table 1.3 are reducible to LLL syllogisms by indirect reduction. I have bracketed the moods that Aristotle does not formulate explicitly.

A second layer of syllogisms combines assertoric propositions with necessity- and/or possibility-propositions. Aristotle accepts the combinations listed in Table 1.4. He takes the LXL moods in Figure 1 to be perfect. The LXL and XLL moods in the other figures he reduces to the first figure by means of modal or non-modal conversion. The remaining moods are equivalent to one or other of these by indirect reduction.

Some of the absences from this table are noteworthy. First, while Aristotle accepts Barbara LXL he rejects Barbara XLL. Supposing that every animal is moving, and given that every man is necessarily an animal, Barbara XLL would imply the falsehood that every man is necessarily moving.[66] This question of the two Barbaras will be a thorn in the side of many a post-Aristotelian logician. Second, Aristotle thinks Baroco XLL and Bocardo LXL are invalid and proposes counter-examples to them. In the case of Baroco XLL his reasoning as it survives in our texts of the *Prior Analytics* is sketchy:[67] he says that a counter-example can be constructed

---

[64] These are equivalent by indirect reduction to the second-figure LLL moods. Aristotle does not assert these moods, however at Aristotle (1989) A22, 40a39-b1 he does assert Datisi LQM.

[65] These are equivalent by indirect reduction to the first-figure LLL moods. The MLM moods are not asserted by Aristotle, however at Aristotle (1989) A22, 40a39-b3 he does assert Disamis and Bocardo QLM.

[66] Aristotle (1989) A9, 30a23-25.

[67] Aristotle (1989) A10, 31a15-17.

using 'the same terms', and if we track that reference back we find the terms 'animal', 'man', 'white'.

It should be mentioned that the syllogisms in Table 1.3 can be seen as derivative on those in Table 1.4, strengthening assertoric premises into necessity-premises – though Aristotle did not see them in this way.

Table 1.4 Aristotle's system: L/X/M combinations

Figure	*Combination*
1	LXL[68]
	[Celarent Ferio LMX[69]; Darii XMM;[70]
	Celarent MLX;[71] Darii Ferio MXM[72]]
2	Cesare Festino LXL;[73] Camestres XLL
	[LMX;[74] Cesare Camestres MLX;[75] Festino MXM[76]]
3	Datisi Ferison LXL, Disamis XLL[77]
	[Ferison LMX;[78] MXM;[79] Datisi Disamis XMM[80]]

68 The first-figure LXL moods are asserted at Aristotle (1989) A9, 30a17-23, 37-b1.

69 These are equivalent by indirect reduction to Festino and Cesare LXL. These moods are not asserted by Aristotle, but at Aristotle (1989) A16, 36a8-15, 34-39 he does assert Celarent and Ferio LQX.

70 This is equivalent by indirect reduction to Camestres XLL. This mood is not asserted by Aristotle, however Darii XQM is provable from Ferison LQX. See Thom (1996) p.71.

71 This is equivalent by indirect reduction to Disamis XLL. This mood is not asserted by Aristotle.

72 These are equivalent by indirect reduction to Ferison and Datisi LXL. These moods are not asserted by Aristotle.

73 The second-figure moods are asserted at Aristotle (1989) A10, 30b9-18 and 31a5-10.

74 These are equivalent by indirect reduction to the first-figure LXL moods. These moods are not asserted by Aristotle, however at Aristotle (1989) A19, 38a16-18, b25-27 he does assert Cesare and Festino LQX.

75 These are equivalent by indirect reduction to Datisi and Ferison LXL. These moods are not asserted by Aristotle, but Camestres QLX is asserted at Aristotle (1989) A19, 38a21-26.

76 This is equivalent by indirect reduction to Disamis XLL. This mood is not asserted by Aristotle.

77 The third-figure moods are asserted at Aristotle (1989) A11, 31b12-20 and 35-37.

The third layer of modal syllogisms consists entirely of contingency-propositions. Aristotle's system is very simple and is shown in Table 1.5. Once again Aristotle takes the first-figure syllogisms to be perfect, the remaining ones reducing to them by means of conversion. Absent from the table are all second-figure moods. This is because of the failure of contingency-conversion for universal negative contingency-propositions. Also noteworthy is the presence of Bocardo QQQ, which, because of complementary conversion is equivalent to Disamis QQQ. (Similarly Darii QQQ is equivalent to Ferio QQQ, and so forth.)

Table 1.5 Aristotle's system: Q/Q combinations

Figure	*Combination*
1	QQQ[81]
2	-
3	QQQ[82]

The fourth layer combines contingency-propositions with assertorics. Again, the first-figure syllogisms are taken to be perfect, and the others reduce by conversion. Again, there are no second-figure moods, for the same reason as with the uniform contingency-moods. These results are displayed in Table 1.6.

Table 1.6 Aristotle's system: X/Q combinations

Figure	*Combination*
1	QXQ[83]
2	-
3	Disamis XQQ; Datisi (Ferison) QXQ[84]

---

[78] This is equivalent by indirect reduction to Camestres XLL. This mood is not asserted by Aristotle, but Ferison LQX is asserted at Aristotle (1989) A22, 40b3-6.

[79] These are equivalent by indirect reduction to the first-figure LXL moods. These moods are not asserted by Aristotle, however at Aristotle (1989) A21, 39b31-39 he does assert Bocardo QXM.

[80] These are equivalent by indirect reduction to Cesare and Festino LXL. These moods are not asserted by Aristotle, however Datisi and Disamis XQM are provable from Bocardo QXM. See Thom (1996) p.76.

[81] The QQQ first-figure moods are asserted at Aristotle (1989) A14, 32b38-a5, 33a21-27.

[82] The third-figure moods are asserted at Aristotle (1989) A20, 39a14-23, 31-38.

[83] The first-figure moods are asserted at Aristotle (1989) A15, 33b33-40, 35a30-35.

The fifth layer combines contingency- and necessity-propositions. The results are displayed in Table 1.7, where again the first-figure moods are said to be perfect, the others reducing to them by means of conversion. The syllogisms in this layer might be seen as derivative on those in Table 1.6, strengthening assertoric premises into necessity-premises – though Aristotle did not see them in this way.

Table 1.7 Aristotle's system: L/Q combinations

Figure	*Combination*
1	QLQ[85]
2	-
3	Disamis LQQ; Datisi (Ferison) QLQ[86]

The sixth layer of syllogisms draws assertoric conclusions from necessity- and contingency-premises. These are shown in Table 1.8. None are perfect. The LQX and QLX moods (apart from Festino LQX) are reduced by Aristotle indirectly to first-figure LXL moods. Festino LQX reduces by modal conversion to Ferio LQX.

Table 1.8 Aristotle's system: L/X/Q combinations

Figure	*Combination*
1	Celarent Ferio LQX[87]
2	LQX;[88] QLX[89]
3	Ferison LQX[90]

[84] The third-figure moods are asserted at Aristotle (1989) A21, 39b26-30.

[85] The first-figure QLQ moods are asserted at Aristotle (1989) A16, 35b23-26.

[86] The third-figure QLQ moods are asserted at Aristotle (1989) A22, 40a16-23, 39-b3.

[87] Aristotle (1989) A16, 36a8-15, 34-39.

[88] Cesare and Festino are proved at Aristotle (1989) A19, 38a16-18; 38b25-27. Camestres can be derived by Aristotelian methods from Ferison QXM: see Thom (1996) pp.61 and 76. Baroco LQX can be proved analogously from Bocardo QXM, which is proved from Barbara LXL at Aristotle (1989) A21, 39b31-39.

[89] Camestres is proved at Aristotle (1989) A19, 38a21-26. Cesare QLX is rejected at Aristotle (1989) A19, 38a38-b5 but can be proved by Aristotelian methods from Darii LXL or Darii QXQ: see Thom (1996) p.129. Baroco QLX can be proved by Aristotelian methods from Barbara QXM: see Thom (1996) p.63. Festino QLX can be proved analogously from Celarent QXQ.

[90] Ferison is proved from Ferio LQX at Aristotle (1989) A22, 40b3-6.

The seventh layer of syllogisms contains a combination of modalities – necessity-, possibility- and/or contingency-propositions, but always with a possibility-conclusion. These are shown in Table 1.9.

There are no perfect moods here. Aristotle demonstrates the validity of Barbara and Darii LQM by indirect reduction to second-figure LLL moods. He demonstrates the validity of Ferio QLM by indirect reduction to Datisi LLL. Second- and third-figure LQM and QLM moods are reduced to the first figure by modal conversion with the exception of Bocardo QLM, which is reduced indirectly to Barbara LLL. Indirect reduction could have been used more widely, to reduce third-figure LQM moods to first-figure LQM moods, and second-figure QLM moods to first-figure QLM moods. All the syllogisms in Table 1.9 could be seen as derivative on those in Table 1.3, strengthening possibility-premises into contingency-premises, the LQM and QLM syllogisms here being consequential upon the LMM and MLM syllogisms there.

Table 1.9 Aristotle's system: L/Q/M combinations

Figure	*Combination*
1	LQM,[91] QLM[92]
2	LQM, QLM[93]
3	LQM, QLM[94]

The eighth and final layer draws possibility-conclusions from assertoric and contingency-premises. Aristotle makes it clear that the assertoric premises in these syllogisms are *simpliciter* [*haplōs*].[95] The procedure whereby he claims to demonstrate the validity of Barbara XQM ('Every *b* is *a*, every *c* may or may not be *b*, so it is possible that every *c* is *a*') is puzzling. He takes the proposition 'It is necessary that some *c* is not *a*', which is incompatible with the conclusion, and adds the assertoric corresponding to the contingency-minor ('every *c* is *b*'), and (using a form of Bocardo) infers that some *b* is not *a* – which is incompatible with the

[91] The first-figure LQM moods are asserted at Aristotle (1989) A16, 35b38-36a2, 15-17, 39-b2.

[92] Aristotle (1989) A16, 36a39-b2.

[93] The second-figure LQM and QLM moods are stated at Aristotle (1989) A19, 38a16-18, 21-26; 38b25-27.

[94] Third-figure LQM and QLM moods: Aristotle (1989) A22, 40a39-b3.

[95] Aristotle (1989) A15, 34b7-8.

original major premise that *a* belongs to all *b*.[96] The puzzlement caused by this procedure can be crystallized into two questions. Why is Aristotle entitled to take the assertoric corresponding to the contingency-minor, rather than the contingency-minor itself? And, what exactly is the Bocardo to which Aristotle thinks he is reducing this Barbara?

Table 1.10 Aristotle's system: X/Q/M combinations

Figure	*Combination*
1	XQM;[97] QXM[98]
2	XQM;[99] QXM[100]
3	XQM;[101] QXM[102]

The syllogisms mentioned in Tables 1.9 and 1.10 could be seen as derivative on those in Table 1.4, strengthening possibility-premises into contingency-premises, the LQX, QLX, QXM, XQM syllogisms here being consequential upon the LMX, MLX, MXM and XMM syllogisms there. Aristotle does not adopt this perspective, but the thought prompts the observation that, since there are some combinations in Tables 1.9 and 1.10 that lack counterparts in Table 1.4, perhaps those absent counterparts would have been accepted by Aristotle. I list the moods in question:

Conjecture 1.1. Barbara, Celarent and Ferio XMM, Barbara MXM, Cesare and Festino XMM, Baroco MLX, Datisi Disamis and Bocardo LMX, and Datisi and Disamis MLX, are valid.

[96] Aristotle (1989) A15, 34a34-b1.

[97] The first-figure XQM moods are asserted at Aristotle (1989) A15, 34a34-b1, 19-27; 35a30-31, 35-b2.

[98] These moods could be seen as derivative on the first-figure QXQ moods. They can also be reduced, by Aristotle's procedure, to third-figure moods.

[99] Cesare and Festino XQM are proved at Aristotle (1989) A18, 37b24-29 and 38a3-4. All the second-figure XQM moods can however be reduced, by Aristotle's procedure, to third-figure moods, Camestres to Ferison and Baroco to Bocardo.

[100] Camestres QXM is proved at Aristotle (1989) A18, 37b24-29. The second-figure moods can be reduced, by Aristotle's procedure, to the first figure.

[101] Datisi and Ferison XQM are proved at Aristotle (1989) A21, 39b26-31. All the third-figure moods can be reduced, by Aristotle's procedure, to first-figure moods, Disamis to Celarent and Bocardo to Barbara.

[102] Disamis QXM is proved at Aristotle (1989) A21, 39b26-31. The third-figure moods can be reduced, by Aristotle's procedure, to the second figure.

Formal analysis

It will be useful to have a uniform and rigorous representation of the various logical theories to be discussed in this book, and to achieve this end I am going to use a formal notation, which I will now explain. I will represent general terms by lower-case letters such as '*a*', '*b*', '*c*', ... I do not presuppose that these terms have existential import; in other words, if the letter '*w*' stands for what is white, I do not imply by the use of this letter that there actually exist any white things. To indicate existential import I will underline the term-letter, so that '$\underline{a}$' will indicate that something is *a*. I shall use an arrow to indicate being-included-in, so that '$b \rightarrow a$' will mean that the *b*s are included in the *a*s. Equivalently, '$a \leftarrow b$' will mean that the *a*s include the *b*s. The backwards arrow then signifies inclusion. I shall use a vertical stroke to signify exclusion, so that '$b \mid a$' means that the *b*s exclude the *a*s. And I shall use the sign '$\smile$' to indicate overlap. With this notation we are able to state truth-conditions for non-modal categorical propositions as in definitions 1.1 to 1.4.

Definition 1.1 (a). 'Every *b* is *a*': $\underline{b} \rightarrow a$.

Definition 1.2 (e). 'No *b* is *a*': $b \mid a$.

Definition 1.3 (i). 'Some *b* is *a*': for some '*d*', $b \leftarrow \underline{d} \rightarrow a$.

Definition 1.4 (o). 'Some *b* is not *a*': for some '*d*', $b \leftarrow d \mid a$.

We also need to explicate the logical relation between '$\smile$' and '$\mid$':

Definition 1.5 ($\smile$). '$b \smile a$' is true iff '$b \mid a$' is not true.

Note that in definition 1.1, '*b*' is required to have existential import, whereas this is not so in definition 1.2 because negative propositions do not have existential import. In Definition 1.3 we need to stipulate that '*d*' is non-empty, otherwise there is no guarantee that some *b* is *a*. However, in definition 1.4 there is no need to do this. 'Some unicorn is not a saint' is true because 'unicorn' is included in 'unicorn' and excludes 'saint'. Clearly we may take any formula of the following types as axiomatic.

Axiom 1.1. If $\underline{b} \rightarrow a$ then $b \smile a$.

Axiom 1.2. If $b \mid a$ then $a \mid b$.

Axiom 1.3. If $c \rightarrow b \rightarrow a$ then $c \rightarrow a$.

Axiom 1.4. If $c \rightarrow b \mid a$ then $c \mid a$.

Axiom 1.5. If $b \nrightarrow a$ then for some '*d*', $b \leftarrow d \mid a$.

Axiom 1.6. If $b \smile a$ then for some '*d*', $b \leftarrow \underline{d} \rightarrow a$.

These last two axioms, in order to be plausible, will need to read as asserting the constructibility rather than the existence in a particular language of a term '*d*' under the stated conditions. Aristotle showed his awareness of this difference when he stipulated that the word 'cloak' stand for 'white man'.[103] Among the theorems we can deduce from these axioms, the following are worth noting.

Theorem 1.1. If for some 'd', $\underline{d} \rightarrow b$ and $\underline{d} \mid a$ then not ($\underline{b} \rightarrow a$). Suppose $\underline{b} \rightarrow a$; then since $\underline{d} \rightarrow b$, we have the chain '$\underline{d} \rightarrow b \rightarrow a$', which simplifies to '$\underline{d} \rightarrow a$' by Axiom 1.3, and this implies '$d \smile a$' by Axiom 1.1, which by Definition 1.5 is inconsistent with our premise '$d \mid a$'.

Theorem 1.2. If for some 'd', $\underline{d} \rightarrow b$ and $\underline{d} \rightarrow a$ then $b \smile a$. Suppose '$b \smile a$' is not true. Then by Definition 1.5, $b \mid a$; so since $\underline{d} \rightarrow a$, we have the chain '$d \rightarrow a \mid b$', which simplifies to '$d \mid b$' by Axiom 1.4, and this is inconsistent with our premise '$\underline{d} \rightarrow b$' by Axiom 1.1 and Definition 1.5. Theorems 1.1 and 1.2 in conjunction with Axioms 1.5 and 1.6, entitle us to express any *particular* (as opposed to universal) proposition in either of two ways – as the denial of an inclusion or exclusion, or else in terms of the existence of a term 'd' linked by inclusion and/or exclusion to the particular proposition's subject and predicate.[104]

Sometimes it will be useful to have a way of representing what necessarily, or possibly, or contingently, falls under a given term. To do this I will superscribe the symbols '*', '†', and '‡', respectively to the term-letter. Thus if 'w' stands for what is white, 'w^*' will stand for what is necessarily white, '$w^\dagger$' for what is possibly white, and '$w^\ddagger$' for what is contingently white (prescinding from the question whether anything is necessarily white etc.). The following axioms govern these notions.

Axiom 1.7. $a^* \rightarrow a$.

Axiom 1.8. $a \rightarrow a^\dagger$.

Axiom 1.9. $a^\ddagger \rightarrow a^\dagger$.

Axiom 1.10. $a^\ddagger \mid a^*$.

The proposition stating that it is necessary that p I will represent as '$\Box\, p$'. The following axioms and rules govern this notion.

Axiom 1.11. If $\Box\, p$ then p.

Axiom 1.12. If $\Box\, b \rightarrow a$ then $\Box\, b^* \rightarrow a^*$.

Axiom 1.13. If $\Box\, b \mid a$ then $\Box\, b^* \mid a^\dagger$.

Axiom 1.14. If $\Box\, b \rightarrow a$ then $\Box\, b^\dagger \rightarrow a^\dagger$.

Axiom 1.15. If $\Box\, b \mid a$ then $\Box\, b^\dagger \mid a^*$.

Axiom 1.16. If $\Box\, b^\dagger \mid a$ then $\Box\, b^\dagger \mid a^\dagger$.[105]

---

[103] Aristotle (1908) Z3, 1029b27-28.

[104] See Aristotle (1989) A28, 44a18-20, 28-30, and Thom (1981) §19.

[105] An analogue of Axioms 1.12 and 1.13 in propositional modal logic is CLCpqLCLpLq, which is a thesis of system T – Hughes and Cresswell (1968) ch.7, A6. An analogue of Axioms 1.14 and 1.15 is CLCpqLCMpMq, which is also

Rule 1.1. If '*p*' is a thesis then '□ *p*' is a thesis.
Rule 1.2. If 'If *p* then *q*' is a thesis then 'If (□ *p*) then (□ *q*)' is a thesis.
Rule 1.3. If 'If *p* and *q* then *r*' is a thesis then 'If (□ *p*) and (□ *q*) then (□ *r*)' is a thesis.

With these notational tools, it will be possible to represent most of the kinds of propositions we will encounter in our survey of medieval modal logic. Moreover, we can string several propositions together in this notation, so that a notation like '$a \rightarrow b \rightarrow b^* \mid c^*$' will mean that whatever is *a* is *b*, and whatever is *b* is necessarily *b*, and nothing that is necessarily *b* is necessarily *c*. Such chains can be transformed in three ways – by shortening, lengthening, or by replacing one link by another. We are entitled to shorten a chain '*c-b-a*' to '*c-a*' when there exists a thesis that if *c-b* and *b-a* then *c-a*; this process I call *simplification*. We are entitled to lengthen a chain '*c-b*' to '*c-b-a*' when there exists a thesis that *b-a*; this process I call *interpolation*. We are entitled to substitute '*d-c*' for '*b-a*' when there exists a thesis that if the former then the latter; this process I call *replacement*.

Essentialist notions

We can define a term '*a*' as being *per se* provided that it is necessary that whatever is *a* is necessarily *a*.
Definition 1.6. '*a*' is *per se* iff $\Box\, a \rightarrow a^*$. 'Horse' is a *per se* term according to Aristotle. 'White' is not a *per se* term, because, even if it happened that all white things were necessarily white, it would still be possible that something white was not necessarily white. To signal that a term is *per se* I shall write it in uppercase. Thus '$C \rightarrow a$' means that the *c*s are among the *a*s and '*c*' is *per se*.
Definition 1.7. '*a*' is *per accidens* iff $\Box\, a \rightarrow a^{\ddagger}$. 'Walking' is a *per accidens* term. In addition to these types, for the sake of completeness, we need to add a third type, which we may call mixed terms '*a*', whose defining characteristic is that some *a*s are necessarily *a* and some are not, e.g., 'white'. (Thus Aristotle's two examples of terms that are not *per se* are of these two distinct types.)
Definition 1.8. '*a*' is mixed iff it is neither *per se* nor *per accidens*. Medieval authors paid scant attention to what I am calling mixed terms. Elsewhere I have called these Quasi-Kind terms.[106]

---

a thesis of T. An analogue of Axiom 1.16 is CLCMpqLCMpLq, which is a thesis of S5 – Hughes and Cresswell (1968) ch.7, TR7.
[106] Thom (1996) p.316.

Besides all this purely logical apparatus, we will need on occasion to make use of some postulates that appear to be assumed in Aristotle's metaphysics. I state three.[107]

Postulate 1.1. If $x \rightarrow a^*$ then for some 'd', $x \rightarrow D$ and $\Box\, d \rightarrow a$. This postulate does not state a formal logical principle, but it seems that Aristotle's essentialism assumes it. What it ensures that whatever is essentially a belongs to some kind that is necessarily included under 'a'. For example anything that is essentially white must belong to a kind that is necessarily included among white things (swans, for example, or snow).

Postulate 1.2. If $x \mid a^\dagger$ then for some 'd', $x \rightarrow D$ and $\Box\, d \mid a^\dagger$.

Postulate 1.3. If $x \rightarrow a^*$ and $\Box\, B \rightarrow a^*$ and $x \nrightarrow B$ then for some 'c', $\Box\, C \rightarrow a$ and $x \rightarrow C$ and $\Box\, C \mid B$. These again are not formal logical principles, but seem to be assumed by two of Aristotle's stated principles: that two intersecting species must both be species of some higher genus,[108] and that whatever is in a genus is in some immediate species of that genus.[109]

[107] Thom (1998) pp.176-177.
[108] Aristotle (1928) Δ2, 121b24-122a2.
[109] Aristotle (1928) Δ2, 122a26-27.

Chapter 2

Late Antiquity

To understand medieval views about necessity and contingency, we need to understand not only the Aristotelian sources on which those medieval theories were based, but also the routes by which the Aristotelian sources were transmitted to the medievals. Aristotle's immediate successors, his Greek commentators, and the Latin translator Boethius, all played distinctive parts in this process of transmission. The transmission traversed a multitude of topics, among them the defensibility of Aristotle's distinction between the two modal Barbaras, the idea of an inseparable accident, the notion that necessities might not be all alike (including the notion that there may be important differences between absolute and conditional necessities), and the notion that the modal concepts of necessity and possibility might themselves be based on some more fundamental concept or concepts.

Theophrastus

Aristotle's pupil Theophrastus, along with Eudemus, is famous for having rejected his master's doctrine about mixed necessity-syllogisms in the first figure. He also abandoned Aristotle's preference for contingency-premises rather than possibility-premises, and made some influential remarks about types of necessity-proposition.

Possibility and contingency

Whereas Aristotle had preferred to consider the premises of modal syllogisms according to whether they were necessity- or contingency-propositions, Theophrastus preferred to consider necessity- and possibility-propositions. This fact explains a number of apparent disagreements between him and Aristotle. For example, whereas Aristotle denies the convertibility of universal negative 'possibility'-propositions Theophrastus accepts it.[1] The dispute is merely apparent, because Aristotle is talking

[1] Alexander (1991) 41, 22 [*ad* A3, 25b19-25].

about Q-propositions while Theophrastus is talking about M-propositions. As noted in Chapter 1, it is a weakness in Aristotle's approach that he does not admit possibility-propositions as premises. For Theophrastus to have done so is, at least potentially, a significant advance. For one thing, it makes it possible in principle to use indirect reduction much more widely than Aristotle does. There is some evidence that Theophrastus did this. For example, Alexander reports that Theophrastus did not prove the validity of Baroco LLL and Bocardo LLL by ecthesis (as Aristotle had done), but instead used indirect reduction.[2] Presumably this means that Theophrastus, unlike Aristotle, explicitly recognized the validity of Barbara LMM and Barbara MLM.

Types of necessity-proposition

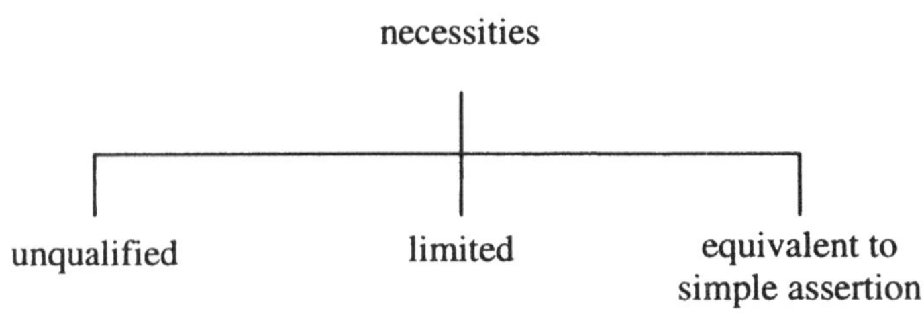

Figure 2.1 Theophrastus's classification of necessities

According to Alexander, Theophrastus distinguished three types of necessity-proposition. First, there are unqualified necessity-statements. These were mentioned in *De Interpretatione* 9, 19a23-26, where Aristotle contrasted them with necessity-statements like 'Everything that is, is by necessity, when it is'. However he did not give a name to this type of necessity-statement, nor did he say whether this was the only type other than unqualified necessity-statements. Theophrastus did both, by opposing unqualified necessity-statements on the one hand to necessity-statements that he called limited [*meta diorismoũ*], and on the other hand to those that express a simple belonging. Alexander states:

> At any rate Theophrastus in the first book of his *Prior Analytics*, discussing the meaning of necessity, writes the following: 'Third, what holds, for when it holds it cannot not hold.'[3]

[2] Alexander (1999a) 123, 19-24 [*ad* A8, 30a2].

[3] Alexander (1991) 36, 25-29 [*ad* A2, 25a9]; Alexander (1999a) 156, 29-157, 2 [*ad* A13, 32a18].

Limited necessity-statements will figure as an important topic in other ancient authors.

The 'peiorem' rule

Aristotle's immediate successors Theophrastus and Eudemus were not convinced by his doctrine that, while Barbara LXL was valid, Barbara XLL was not. Their rule – known in the Middle Ages as the *peiorem* rule – was that the conclusion should follow the weakest premise. This rule was meant to encompass three different aspects of categorical propositions – their quantity, quality and modality. In respect of quantity, the idea was that universality is stronger than particularity (presumably because while a particular proposition follows from a universal, a universal never follows from a particular); accordingly, a universal conclusion can be deduced only from universal premises. If there is a particular premise, the conclusion must be particular. This result is standard Aristotelian doctrine. In respect of quality, affirmatives are stronger than negatives. An analogous reason can be supplied: while a negative follows from an affirmative, according to Aristotle an affirmative never follows from a negative. Here, the Aristotelian doctrine in non-modal syllogistic was that an affirmative conclusion follows only from premises all of which are affirmative. If there is a negative premise, the conclusion too must be negative. Extending this pattern to modality, Theophrastus and Eudemus thought that since necessity-propositions are stronger than assertorics (again because, while a statement of fact follows from one of necessity, the reverse never happens), the Aristotelian position *should* be that a necessity-conclusion follows only from premises all of which are necessity-propositions, and if there is an assertoric premise the conclusion too should be assertoric. By contrast, what Aristotle actually says is that if one of the premises (the minor) is assertoric a necessity-conclusion follows, provided that the other premise is a necessity-proposition.[4]

The theory of Theophrastus and Eudemus constitutes a clear and a principled challenge to Aristotle's authority in modal syllogistic. As such it would become one of the main topics of discussion amongst medieval modal logicians. That said, their theory has some obvious shortcomings. First, it is not clear how the *peiorem* rule is supposed to apply to contingency-propositions: are they supposed to be stronger or weaker than necessity-propositions or assertorics? Second, the *peiorem* rule appears to conflict with Theophrastus's widespread use of indirect reduction, because

[4] See Alexander (1999a) 124, 1ff [*ad* A9, 30a15].

if two syllogisms are equivalent by indirect reduction and one conforms with the *peiorem* rule, the other may not. For example, Barbara LXL (which does not conform with the *peiorem* rule) is equivalent with Bocardo MXM (which does conform). The Theophrastus property (as we may call it) – that the conclusion is no stronger than the weakest premise – is not preserved by the procedure of indirect reduction.

Sosigenes

Alexander's teacher Sosigenes is reputed to have believed that Barbara LXL, if it is to be valid, should be understood as having an unconditional necessity-major, and a *conditional* necessity-conclusion. We may take it then that the syllogism would be along the lines of 'All humans are necessarily animals, everything moving is human, so everything moving is necessarily an animal *so long as it is human*'.[5]

From one point of view, Sosigenes might extract our concurrence on this point without much demur. After all, *any* premises entail the conclusion that everything moving is an animal so long as it is human. The reason is that *this* conclusion – as distinct from the proposition that everything moving is an animal – is necessary relative only to the existence of humans, and therefore follows from absolutely anything (given that there are humans). In addition to this rather obvious claim, Sosigenes's thesis is clearly meant to imply that an *un*conditional conclusion does not follow from those premises. In fact, Sosigenes's thesis is consistent with Theophrastus's view that an unconditional necessity *never* follows from premises one of which is non-modal. Some things (some primary substances) that are not moving at a particular time may be moving at another time. So, it cannot be absolutely necessary that everything moving is human. Given this, the major premise that it is absolutely necessary for humans to be animals, together with the minor premise that everything moving is human cannot imply a conclusion that it is absolutely necessary that everything moving is an animal.

What will be influential is Sosigenes's use of conditional necessity-statements. Such statements will receive attention from Boethius and Abélard among others.

---

[5] Alexander (1999b) Appendix 3.

Alexander of Aphrodisias

Alexander (fl. c.193-217) wrote a commentary on the *Prior Analytics* which, apart from expounding the Aristotelian text, contains many passages that frustrate the reader in search of theory and system, because they outline arguments on both sides of a question without finally endorsing one or the other. I shall concentrate on what Alexander has to say about the notion of possibility, inseparable accidents, types of necessity-proposition, necessity-syllogisms, and contingency-syllogisms.

Possibility

Alexander describes three accounts of possibility – Diodorus's definition of the possible as what either is or will be, Philo's definition of the possible as 'the suitable for the subject, even if it has been prevented from coming to be by some external necessity', and Aristotle's account.[6] On Philo's account, says Alexander, it is possible for chaff that is lying in unmown wheat or is at the bottom of the ocean to be burnt. Alexander thinks that Aristotle's understanding of possibility is intermediate between the Diodorean and Philonian accounts because on the one hand Aristotle allows for possibilities that will not be realized, and on the other hand Aristotle does not think that the possibilities for individual members of a species are unaffected by accidents that happen to them (such as being thrown into the sea). According to Aristotle, such an accident may close off possibilities for an individual.

Be this as it may, Alexander is not very forthcoming about what exactly Aristotle's account of possibility is. It may be desirable to have a concept of possibility that allows for possibilities that will not be realized, and recognizes that the possibilities for individual members of a species are affected by accidents that happen to them; but Alexander does not articulate a theory of possibility that achieves these aims. We will find the beginnings of such a theory in Boethius.

Inseparable accidents

Alexander, commenting on *Topics* A5, takes the line that whiteness is not of the essence of snow but rather is accidental to it, even though it is necessary that snow is white. And his argument for the necessity of snow's whiteness is worth quoting:

---

[6] Alexander (1999b) 184, 1-19 [*ad* A15, 34a12].

> Thus a certain coagulation of a certain type of body – e.g. of cloud, which is the material of snow – generated by excessive cold and accompanied by an internal deficiency of air, is necessarily followed by whiteness: the cause of such a colour is a deficiency of the air within the cloud, the coagulation of the cloud and the friction of the air which in this process is caused by excessive cold and prevents the air being separated off. For air becomes white by being rubbed and stirred – and rubbing and stirring occur in the coagulation of the cloud, as the air within it is condensed and shut away into its depth and hindered from being separated off. These other features belong to the essence of snow, but the whiteness attending them is a modification and an accidental of it. It would seem to be an accidental of it, because it is not something that brings to fulfillment its essence: for if white is neither the matter nor the form of snow, it is not in the essence of snow; it becomes inseparable from it as an epiphenomenon of what happens to its matter in its change into such a form as we have described.[7]

Alexander thus commits himself to the existence of necessary accidents. On the face of it, he is at odds with Porphyry on this question, since Porphyry's view is that whiteness can be imagined away from the subjects for which Aristotle had said it was necessary. We shall explore the consequences of these opposed views shortly. In the meantime, we should also note that the grounds on which Alexander and Porphyry rest their statements are quite different from one another, with Alexander relying on an argument drawn from physical necessities while Porphyry relies on a thought-experiment. Because of this difference it might be argued that the two views can be reconciled. Given the physical laws that in fact govern the world, whiteness is necessary for snow, but those laws do not apply in all conceivable worlds, and were they not to apply snow might not be white.

Types of necessity-proposition

Alexander gives a couple of examples that will become objects of study in the medieval period. As an example of a limited necessity Alexander mentions the proposition that of necessity everything literate is human (sc. so long as it is literate). He says that this is not an unconditional necessity:

> Theophrastus showed the difference...; for there are not always literate things, and a human is not always literate.[8]

---

[7] Alexander (2001) 50, 20 - 51, 4 [*ad* A5, 102b4-14].

[8] Alexander (1991) 36, 25-31 [*ad* A3, 25a29-31].

Alexander goes on to say that when Aristotle speaks of the convertibility of necessity-propositions he means unconditional necessity-propositions. This example will assume crucial importance in the thinking of Robert Kilwardby.

Again, Alexander thinks that it is not necessary in an unconditional way that whatever is walking is moving:

> For all that walks does not move by necessity if, indeed, it is true that what walks does not even walk by necessity except, as I said, on the condition 'as long as it is walking'.[9]

This example will assume special significance for Averröes.

Alexander states that Aristotle's discussion of apodeictic conversion concerns apodeictics that are not conditional in this way. By implication Alexander is saying that modals with such a temporal limitation do not convert like assertorics – a view that will be revived by Kilwardby.

Mixed necessity-syllogisms

Our knowledge of the *peiorem* rule comes mainly from Alexander's commentary on the *Prior Analytics*. In that commentary, he outlines a number of arguments for and against Theophrastus's position. First, there is the Theophrastian argument that I outlined earlier. Alexander does not subject the argument to much scrutiny – doubtless, as he himself says,[10] because he had discussed it at length in his (lost) work *On the difference between Aristotle and his successors concerning mixtures of propositions* – but it can be seen as resting on two premises. One premise posits an analogy among the quantitative, qualitative and modal features of a categorical proposition: in each case the range of features in question can be ordered so that each pair of features in the range is such that a proposition possessing the first feature can imply one possessing the second but not vice versa. (An affirmative can imply a negative but not vice versa, and so on.) The first feature is then said to be *stronger* than the second. Let us say that when each feature in a set is either stronger or weaker than every other, then they are all *comparable*. The other premise in Theophrastus's argument asserts that when the premises and conclusion of a syllogism are all comparable, then the conclusion cannot be stronger than any premise. Now, the first premise is unassailable, but the second is questionable. The fact that, say, a universal proposition never follows from

[9] Alexander (1999a) 155, 23-25 [*ad* A12, 32a12].

[10] Alexander (1999a) 125, 30-31 [*ad* A9, 30a15].

a single particular does not obviously entail that a universal cannot be deduced from a set of propositions one of which is particular.

Alexander mentions a second argument, namely that if *b* holds of all *c* but not by necessity, then it is possible that *b* sometimes be disjoined from *c*, but in that case it is possible that *a* sometimes be disjoined from *c*.[11] This is not fully clear, and Alexander again does not pursue the matter; but if by 'it is possible that *b* is sometimes disjoined from *c*', he means either that some *c* is possibly not *b*, or that some *b* is possibly not *c*, then this argument either assumes the validity of the inference '*a* is necessary for all *b*, some *c* is possibly not *b*, so some *c* is possibly not *a*' (LaMoMo-1), or else the validity of '*a* is necessary for all *b*, some *b* is possibly not *c*, so some *c* is possibly not *a*' (LaMoMo-3). These two moods are equivalent to LaLaLa-2, an obviously invalid mood.

Against the validity of Barbara LXL, Alexander mentions three putative counter-examples. These have the terms 'animal' 'human' 'moving', 'having knowledge' 'literate' 'human', and 'moving' 'walking' 'human'.[12] He reports a view according to which Aristotle means that the assertoric minor premise in Barbara LXL is actually asserted universally, not just assumed for the sake of argument. He points out that the minor premises in the putative counter-examples cannot be asserted as true universally, and states that, according to this view:

> The conclusion is not necessary when the premisses are taken as holding in this way, because they are not true when taken as holding universally.[13]

Alexander appears to making two points – first, that the minor premises in the alleged counter-examples ('Everything moving is human', 'All humans are literate' and 'All humans are walking') could at best be true for a limited time, and second that the conclusions ('Everything moving is necessarily an animal', 'All humans are necessarily knowledgeable', and 'All humans are necessarily moving') are not true or at best would be true only under certain conditions. What he appears to be suggesting is that the minor premise in Barbara LXL must be taken *simpliciter* and not merely *ut nunc*; and this is a view that will command many adherents in the Middle Ages. He also raises a difficulty for this view. How, he asks, will it deal

---

[11] Alexander (1999a) 124, 18ff [*ad* A9, 30a15].

[12] Alexander (1999a) 124, 21-31 [*ad* A9, 30a15]. Mueller and Gould mention that Philoponus uses the trios 'moving' 'walking' 'human' and 'virtue' 'practical wisdom' 'human'.

[13] Alexander (1999a) 126, 15 [*ad* A9, 30a15].

with Darii and Ferio LXL, where the assertoric minor premise is particular, not universal. The point is that:

> ...it is no longer possible to say that the particular unqualified premiss is not true, for example, 'Human holds of something that is moving' or 'Literate holds of some human'.[14]

In other words, what are the truth-conditions for particular *simpliciter* assertorics?

In favour of the validity of Barbara LXL, Alexander mentions its equivalence to Bocardo MXM – a mood that both Aristotle and Theophrastus ought to accept.[15]

The two Barbaras are not the only disputed moods in Aristotle's account of mixed necessity-syllogisms. Controversy has also surrounded Baroco XLL and Bocardo LXL. Baroco XLL is a mood that Aristotle rejects as invalid. We have seen that his counter-example is unconvincing, and Alexander is aware of this.[16] He reads the text as saying 'the same terms if just one of them is changed', and he searches for a suitable group of terms, trying the trios 'awake' 'literate' 'human', 'animal' 'moving' 'white', and 'biped' 'awake' 'animal'. He seems to prefer the second trio. Those terms generate the argument 'Animal belongs to everything moving, animal is impossible for some white, so movement is impossible for some white', and here we can perhaps see how the premises can be true while the conclusion is false. Animal is impossible for some white, by virtue of the fact that snow is white, and we may suppose that only animals are in fact moving, but it seems clear that these premises do not entail that something white is incapable of movement. So it seems that he has produced a counter-example to Baroco XLL. On the other hand, Alexander sees that Baroco XLL appears to be provable by just the kind of ecthetic proof that Aristotle used to prove Baroco LLL.[17] If every *b* is *a* and some *c* cannot be *a*, then we can take some of *c* – say, *d* – that cannot be *a*, and reason with regard to it: no *d* can be *a*, but every *b* is *a*, so no *d* can be *b* (Camestres XLL); and from this it follows that some *c* cannot be *b*. So Alexander has outlined a difficulty, by presenting arguments both for and against the validity of Baroco XLL. This difficulty will excite the attention of several medieval logicians. Among those who will accept the validity of Baroco XLL are Avicenna, Campsall, Ockham and Buridan. We will find Robert

---

[14] Alexander (1999a) 126, 21 [*ad* A9, 30a15].
[15] Alexander (1999a) 127, 3ff [*ad* A9, 30a15].
[16] Alexander (1999a) 144, 4-23 [*ad* A10, 31a10].
[17] Alexander (1999a) 144, 24f [*ad* A10, 31a10].

Kilwardby occupying the embarrassing position of denying its validity while showing us how to prove it.

Alexander also raises a doubt about the invalidity of Bocardo LXL. He mentions a textual uncertainty: some manuscripts make the middle-term 'biped' not 'animal' as it clearly should be. However he accepts Aristotle's counter-example, while noting that the mood's validity seems to be provable by an ecthetic proof analogous to the one Aristotle uses for Bocardo LLL.[18] I do not read his commentary here as endorsing the Aristotelian rejection of this mood, but rather as formulating a difficulty by setting out the arguments on both sides of the question. This mood will also find its medieval supporters, including Abélard, Avicenna, Averröes and Ockham.

Contingency-syllogisms

Alexander, commenting on Aristotle's distinction between the ampliated and unampliated senses of contingency-propositions, remarks:

> Consequently 'It is contingent that A holds of all B' can also mean two things. But when the proposition means the first one mentioned [sc. the unampliated sense], the premisses are not both contingent; rather the minor is unqualified, the major contingent, and such a combination is mixed. When the proposition means the second [sc. the ampliated sense], both premisses are contingent.[19]

It appears from this that Alexander thinks that the meaning of contingency-propositions in the mixed contingency-syllogistic is different from that in the pure system. In the former case contingency-propositions are unampliated, in the latter ampliated.[20]

In several places, Alexander uses indirect reduction to demonstrate the validity of mixed contingency-syllogisms that Aristotle had either ignored or rejected as invalid. Camestres LQM is one such. Aristotle does not mention it, but Alexander shows that it is valid by reducing it indirectly to Darii LLL:[21]

> I was perplexed about why these combinations having one premiss universal affirmative necessary and the other contingent universal negative are not syllogistic. For it seems possible for a contingent universal negative conclusion

[18] Alexander (1999a) 150, 25-151, 30 [*ad* A11, 32a4].
[19] Alexander (1999a) 166, 5ff [*ad* A13, 32b29].
[20] See Patterson (1995) p.141ff, Thom (1996) p.36ff. But compare Alexander (1999a) p.139f.
[21] See Thom (1996) p.73 for details.

> to be proved by *reductio ad impossibile*. For if it is assumed that A holds of all B by necessity and that it is contingent that it holds of no C, I say that it is contingent [*sc.* 'possible'] that B holds of no C. For if not, it is not contingent that it holds of no C, i.e., it holds of some C by necessity. But A holds of all B by necessity. Therefore, A holds of some C by necessity, which is impossible, since it was assumed that it is contingent that A holds of none of it.[22]

This example exemplifies Alexander's tendency to make far wider use than Aristotle did of the procedure of indirect reduction (*reductio ad impossibile*). Here he proves the validity of Camestres LQM (or Camestres LMM) by reducing it to that of Darii LLL. This reduction is correct. What Alexander is doing here is filling in one of Aristotle's omissions in Table 1.3. This tendency will be carried to its ultimate point eventually by Buridan.

Aristotle thinks all moods deriving a universal negative necessity-conclusion from a universal negative necessity-major and a contingency-minor in the first figure are invalid.[23] This commits him to rejecting Celarent LQL (though he accepts Celarent LQX). Alexander disagrees, and proposes to deduce the validity of Celarent LQL from that of Disamis MQM (or MMM).[24] Aware that he is in conflict with Aristotle, he states:

> It is worth asking why he says that in a mixture of a necessary negative universal premiss and a contingent affirmative one there is no necessary negative conclusion that X holds of no Y by necessity in any of the figures. For either (i) it is necessary that *reductio ad impossibile* be rejected; or (ii) the combinations in the third figure through which I produced the *reductio ad impossibile* [sc. in this instance Disamis MMM] must be non-syllogistic; or (iii) it follows that X holds of no Y by necessity [sc. and thus Celarent LQL is valid]. ...I have investigated this and spoken about it at greater length in *On the disagreement of Aristotle and his associates concerning mixtures*. And I have discussed it at greater length in my notes on logic.[25]

Unfortunately, both the works to which Alexander refers here are lost. What we can say, however, is that Alexander's *reductio* is correctly performed: Celarent LML (or LQL) does indeed follow from Disamis MMM (or MQM).[26] However, it is not obvious that Disamis MMM is valid. If affirmative possibility-propositions state simply that their

---

[22] Alexander (1999b) 238, 22-29 [*ad* A19, 38b4].
[23] Aristotle (1989) A16, 35b34-35.
[24] Alexander (1999b) 249, 18-24, [*ad* A22, 40a9].
[25] Alexander (1999b) 249, 32-250, 2 [*ad* A22, 40a9].
[26] See Thom (1996) p.58 n.9.

predicates are possibly true of their subjects then Disamis MMM is not valid, since it makes the inference '*a* is possible for some *c*, *b* is possible for all *c*, so *a* is possible for some *b*' when the most that can be inferred is that *a* is possible for something for which *b* is possible. It would be different if affirmative possibility-propositions were understood as stating that their predicates are possibly true of what their subjects are possibly true of: in that case Disamis MMM would be valid. We shall return to this question as we examine the systems of Ockham and Buridan.

Alexander notices that some second-figure LQM moods are indirectly reducible to LLL moods in the first figure. He cites the syllogism with universal affirmative necessity-major, and universal contingency-minor, as being reducible to Darii LLL.[27] He does not say so, but this method of proof can be generalized to all second-figure LQM moods.

Porphyry

Porphyry (c.232-305) is best known for his much-commented *Introduction to Aristotle's Categories*, which deals with the doctrine of the predicables and develops what Aristotle had to say on this topic in several important ways, especially in relation to peculiarities and accidents.

Peculiarities

Porphyry states that his predecessors (unnamed) distinguish four types of peculiarity. (1) The first type are features, like healing and measuring in man, that are confined to one species but do not occur in every member of the species.[28] These activities are exclusive to humankind, and necessarily so. Nothing that was not human could engage in healing or in measuring – not as an intentional activity, in the way typified by human doctors and surveyors. On the other hand, it is not necessary but contingent that a particular man engages in healing or measuring. In Aristotle's terminology, these features are accidental to man and not convertible, so Aristotle himself would have called them accidents rather than peculiarities. (2) The second type of peculiarity comprises features such as two-footedness that occur in every member of the species but also in other species. This is a necessary characteristic of all humans, but also belongs with necessity to other species such as various types of bird. Again, these do not convert with the species, so Aristotle would not have regarded them as peculiarities.

[27] Alexander (1999b) 240, 4-11 [*ad* A19, 38b13].
[28] Porphyry (1975) 12:12.

(3) Thirdly there are peculiarities that are confined to a single species and occur in every member of that species, but not all the time. Porphyry cites the fact that humans go grey. Recall that going grey had been nominated by Aristotle as an instance of natural contingency. Porphyry seems to think, all the same, that this property is confined to the human species. (Maybe he had not seen old dogs, but let that pass!) In any case, humans do not go grey *all the time* but only in old age. We should distinguish here between actually going grey and having the capacity to go grey. The former Aristotle would have regarded as an accident, the latter as a peculiarity of those species having the capacity in question. (4) Finally and most properly, some peculiarities occur only in one species, in every member of the species, and all the time. His example is the capacity for laughter in humans. He believes that this capacity belongs to the human species as a matter of necessity. Individual members of the species may lack this capacity; but that very fact would mean they were deficient as members of the species.

We will encounter some of these types of peculiarity repeatedly as we traverse the history of medieval modal logic. The first type – exemplified in Porphyry by the activities of healing and measuring – will recur in the thirteenth century in the form of discussion about the capacity for *literacy* among humans. The second type – exemplified in Porphyry via the feature of two-footedness in humans – will recur in discussions of the whiteness of snow and the blackness of Ethiopians. These types, along with the fourth – the capacity for laughter – are standardly taken as belonging *necessarily* to their subjects. The third type – the tendency to go grey – will occur more in discussions of contingency rather than necessity.

Inseparable accidents

Aristotle repeatedly affirms in the *Prior Analytics* that snow and swans are white not just as a matter of fact but necessarily, and that pitch and ravens are black not just as it happens but of necessity. In the face of these examples, an Aristotelian scholar may well be provoked to ask, does Aristotle really think that a colour such as whiteness or blackness is *essential* to any subject? Does he not classify colours in the *Categories* as accidents rather than essential features? If he is right about all that, then it seems that at least some individual subjects possess features that belong to them necessarily but not essentially. In other words there exist necessary accidents. But then, Aristotle in the *Topics* speaks of accidents as if they were non-necessary features of their subjects, when he describes an accident as something that can either belong or not belong to its subject.

It would seem as though the views expressed in the *Categories*, the *Analytics* and the *Topics* do not sit happily together.

In his *Introduction to the Categories* Porphyry discusses inseparable accidents in two places, once in the section on the differentia, and then again in the section on accidents. In the differentia section, he distinguishes three meanings of the word 'difference'. The *common* sense is the broadest; in this sense things differ when there is any respect in which they are not the same, for example, Socrates differs from Plato, and the old Socrates differs from the young. The *proper* sense concerns individual differences arising from an inseparable accident. Porphyry's examples are greyness [*glaukotēs*] of the eyes, a hooked quality of the nose, or 'a scar which has become hardened from a wound'. In the *strict* sense a difference is a specific difference such as rationality for humans.[29] Porphyry goes on to state that the first two types of difference are qualitative, not essential – even (we may add) if they are inseparable.

This passage is interesting, because it concerns what is necessary or impossible for an individual, and because the modality in question is a temporal one: it was not inevitable that this man should have a scar, but once he is scarred he will always be scarred, and in that sense it is impossible that he will (after that) be unscarred.

In the section on accidents, Porphyry considers the two definitions of an accident that are found in Aristotle's *Topics* – 'what belongs to a subject but is neither genus nor difference nor species nor peculiarity' and 'what can belong and not belong to the same thing [*tōi autōi*']'. The latter definition appears to exclude the possibility of inseparable accidents; however, Porphyry understands it to mean, 'what can belong and not belong to the same substratum [*hypokeimenou*]'. The distinction is vital in his consideration of his examples (the blackness of crows and Ethiopians).[30] He states that even though blackness is inseparable from all actual crows and Ethiopians 'it is possible, however, to conceive of a white crow and of an Ethiopian who has lost his color apart from the destruction of the substratum'.[31] This thought will be taken up by Porphyry's medieval successors.

Porphyry's discussion of class-wide inseparable accidents is interesting, and for two reasons. The first point of interest concerns the Ethiopian example. Crows form an Aristotelian species, being a natural kind; but although Ethiopians form an identifiable class of humans, they do not form

---

[29] Porphyry (1975) 8:8-18.

[30] Porphyry later adds charcoal and ebony as further examples of classes from which blackness is inseparable. Porphyry (1975) 22:5.

[31] Porphyry (1975) 13:1.

a sub-species of humankind. There are, according to Aristotle, no such sub-species. So peculiarities, in Porphyry's eyes, belong to classes, not only to species.

Second, Porphyry's view that although in reality blackness cannot be separated from the classes he mentions, we could imagine it being separated, implies a contrast with genus-species necessities, where we could not even imagine a separation. Perhaps we might be able to capture this difference by referring, not just to actual, but also to all possible crows or Ethiopians. A genus belongs inseparably not only to all actual members of its species but to all possible members. However, an inseparable accident, while belonging inseparably to all actual members of the class in question, may not belong to all possible members. Some possible crows and Ethiopians are not black, even if no actual ones are. This distinction will take on considerable importance in the modal logic of Buridan.

Ammonius

In the commentaries of Ammonius (fl. c.500) we find further development of the idea of qualified necessity-statements. Ammonius, commenting on *De Interpretatione* 19a23, distinguishes propositions of absolute necessity ('Of necessity the Sun moves', 'Of necessity the angles of a triangle are equal to two right angles', 'Of necessity this fire is hot', 'Of necessity Socrates is an animal') from ones that contain the qualification [*prosdiorismos*] 'as long as that is true which is predicated by the one who says that it is so' ('The Sun is of necessity obscured by a cloud, or by the Moon, as long as it is obscured' or 'You are of necessity walking, or sitting, as long as you are walking, or sitting'). Propositions of absolute necessity, so understood, are of two types, depending on whether they indicate an eternal truth or a truth so long as the subject exists.[32] The notion of hypothetical necessity will play an important role in Avicenna's modal logic.

Ammonius's threefold distinction is again found in Stephanus:

> ...there is what is necessary absolutely and what is necessary hypothetically. We say that something is necessary hypothetically, so long as the predicate belongs to the subject. But that same thing is capable also of not belonging, as that Socrates should be sitting or sleeping. So long as the sleeper sleeps, of necessity sleep belongs to him. The absolutely necessary is again twofold: it exists both in the case of eternal things and in the case of things that come to be

[32] Ammonius (1998) p.115.

and cease to be, when a substantial differentia of the thing that is subject is present. We say 'substantial' because in the case of the hypothetically necessary when the thing, that is, the predicate, is separated, for instance walking or sleeping, the subject is not destroyed, but here, for instance in the case of fire if the heat is separated it destroys the subject, that is, the fire. What is the necessity that is to be seen in the case of eternal things? For example, when we say 'The Sun is of necessity in motion', 'God is of necessity good.' But in the case of things that come to be and cease to be, it is like this fire. Like the particular, we mean, and not the universal, since this particular fire comes to be and ceases to be, but the universal is always the same and does not cease to be.[33]

Boethius

Inseparable accidents

Boethius (c.480-c.524) translates Porphyry's Alexander's term '*glaukotēs*' as blindness [*caecitas*], thereby putting blindness into the category of an inseparable accident of those subjects that it characterizes. In so doing, Boethius unintentionally set going a debate about whether those who are blind are necessarily so.

Repugnance

One way of articulating a notion of possibility that allows for possibilities that will not be realized, and recognizes that the possibilities for individual members of a species are affected by accidents that happen to them would be via a notion that played an important role in some medieval writings on logic, the notion of repugnancy. What I have in mind is the idea that the possibilities for a subject consist in what is not repugnant to certain of that subject's features. Boethius treats *repugnantia* (Stump translates 'incompatibles') in different ways in different passages, and in some cases confusedly. We can distinguish four different definitions. Stump explains one definition as follows:

> ...Boethius understands incompatibles in this way. If x and y are contraries and if z entails y or y entails z, then x and z are incompatibles.[34]

Stump does not observe – nor does Boethius – that under this definition x and z may be called incompatibles when they can be true together, indeed

[33] Stephanus (2000) p.158 [*ad* 19a23-24].

[34] Boethius (1988) p.225 n.4.

when one entails the other. To see this, let *z* be *x or y*. The trouble arises from the clause 'or *y* entails *z*'. If that clause were omitted we would have a definition of a particular type of incompatibility, namely where one of the incompatibles is subordinate to one of a pair of contraries, and the other incompatible is the other contrary. This is indeed a type of incompatibility, because if *x* and *y* are contraries, and *z* entails *y*, then *z* and *x* are incompatible. By contrast, if *x* and *y* are contraries, and *y* entails *z*, then it does not follow that *z* and *x* are contraries.

Later, Stump comments on a second definition:

> In the earlier discussion the incompatibles used in the examples are properties or terms. Here [Book V, 354/1131, Stump p.133] it is propositions that are said to be incompatible, and the general criteria for such incompatibles seem to be this: if *p* entails *q* and if we negate *q*, then '*p* entails ~*q*' is an incompatible.[35]

In line with this second definition Boethius writes:

> An incompatible is that which cannot exist simultaneously with the thing with which it is said to be incompatible.[36]

His examples are 'day'-'not light' and 'animal'-'dead'. Elsewhere he puts forward a third definition, which was to be influential in the Middle Ages:

> ...incompatibles are not themselves contraries but rather the consequents of contraries (as we showed in our exposition of the Topics presented by Cicero).[37]

He offers an example:

> ...since friendship and enmity are contraries and a will to injure follows from enmity, friendship and a will to injure are incompatibles. These things are also related to one another in the manner of contrariety.[38]

On this definition, as on the first, it turns out that what he calls incompatibles need not in the logical sense be incompatible with one another. To think otherwise is to commit the logical fallacy of inferring 'No *f* is *w*' from the premises 'No *f* is *e*' and 'Every *e* is *w*'.

---

[35] *Ibid.*

[36] Boethius (1988) 347/1124, p.124.

[37] Boethius (1988) 350/1127, p.127.

[38] Boethius (1988) 295/1066, p.52.

Fourthly, we find in some of his examples an understanding of incompatibles, not as the consequents but rather as the antecedents of contraries – on which definition incompatibles are indeed incompatible. Among these examples there is one that will be taken up by the medievals – 'waking' and 'snoring'.[39] The argument here would be 'Nothing that is awake is asleep, and everything that is snoring is asleep, therefore nothing that is awake is snoring' (Camestres).

Types of necessity-proposition

Anyone wishing to theorize the notion of necessity has to face the question whether necessities are all alike or whether there are interestingly different types of necessity. Boethius gives different classifications of necessities in different places. In *De hypotheticis syllogismis* he essentially follows Theophrastus's tripartite division, though he puts them in the reverse order; and unlike Theophrastus he presents a tripartite division of possibilities in parallel with that of necessities. He divides statements of necessity and possibility into three species: necessity is either equivalent to simple assertion ('Socrates necessarily sits so longs as he sits' = 'Socrates sits') or conditional ('Socrates necessarily has a heart so long as he lives') or absolute ('God is immortal'); and possibility is either equivalent to a statement of fact ('Socrates possibly sits, while he sits') or conditional ('Socrates possibly reads, so long as he is Socrates') or absolute ('A bird possibly flies').[40] The division (like that of Theophrastus) seems flawed since the second type appears to include the first: any statement like 'Socrates necessarily sits so long as he sits' is going to be *both* conditional *and* equivalent to a simple assertion. It appears, in other words, that Boethius has divided necessities into the absolute and the conditional, identifying a sub-class of the latter as being equivalent to simple assertions.

In his first commentary on *De Interpretatione* Boethius distinguishes propositions of temporal necessity ('Being seated is necessary for me when I am seated') from propositions of unconditional necessity ('Mortality is necessary for a human being'). The first type corresponds to his 'necessity-propositions that are equivalent to simple assertions', and thus is a conditional necessity. The second type, while described as unconditional, does not seem to be quite as unconditional as his example 'God is immortal', since while there is no time at which God does not exist there may well be a time at which no human being exists, and to this extent the

---

[39] Boethius (1988) 354/1131, p.134.
[40] Boethius (1969) I, VI, 6-7, pp.236-238.

necessity of the statement that mortality is necessary for a human being seems to be conditional on the existence of a human being.

This point is taken up in the second commentary on *De Interpretatione*, where Boethius makes a different distinction, that between necessity-propositions expressed as 'simple predications' ('It is necessary that the sun move') and those that are 'put forward with the necessity of some accidental characteristic' ('It is necessary that Socrates is seated when he is seated').[41] This appears to be the same as his distinction in *De hypotheticis syllogismis* between absolute and conditional necessity, and indeed he describes these two types of necessity as unconditional and conditioned, pointing out that the former does not entail the latter; and he gives a further example of unconditional necessity:

> But if we say of a marble Socrates that it is necessary that that marble Socrates be seated (if it happens to have been carved seated), it is true; and necessity could be predicated unconditionally of such a Socrates.[42]

The marble Socrates, of course, will cease to exist sometime; so the necessity of its being seated should be relativized to its continued existence. Boethius seems here to be following Ammonius's doctrine that 'absolute' necessities are conditional on the existence of their subjects.

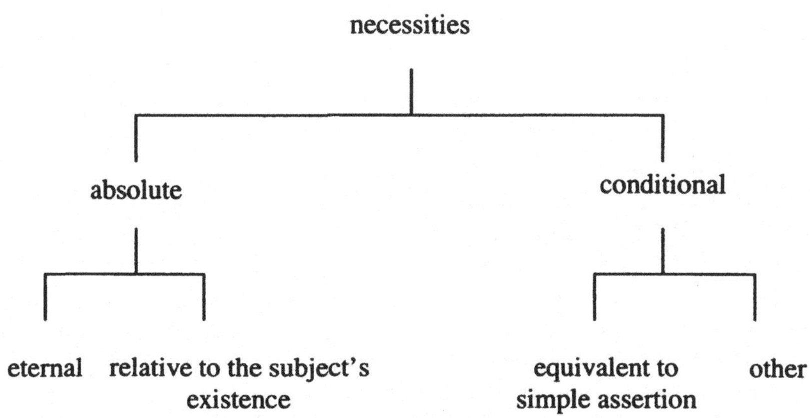

Figure 2.2 Boethius's classification of necessities

In his second commentary on *De Interpretatione* (Introduction to ch.9) Boethius distinguishes propositions that are *immutably necessary* ('God

[41] In Ammonius (1998) 241, 1 [*ad* 19a23].
[42] In Ammonius (1998) 242, 25 [*ad* 19a23].

exists' or 'God is immortal' or 'The number of the stars is either even or odd') from ones that are *necessary to the extent to which the subjects the proposition affirms or denies something about do exist* ('A human being is mortal' or 'Fire is hot').[43] These are the two types of absolute necessity-proposition distinguished by Ammonius. Overall, then, we can attribute to Boethius a division of necessity-propositions as shown in Figure 2.2.

Formal analysis

Inseparable accidents

The notation introduced in Chapter 1 can be used to provide a formal analysis of Alexander's and Porphyry's distinct positions in relation to inseparable accidents. If we take a Kind '*k*', and any term '*a*', '*a*' will be an inseparable accident of '*k*' if and only if $\Box\, K \rightarrow a$ (where '*a*' is not a Kind-term). Relying in Definition 1.6, we can say that '*a*' is a necessary accident of '*k*' provided that (1) $\Box\, k \rightarrow k^*$, (2) $\Box\, k^* \rightarrow a$ and (3) $\sim\Box\, a \rightarrow a^*$. Thus, the anti-Alexandrian thesis, according to which there are no necessary accidents, can be represented as stating that if (1) and (2) then not (3). This thesis states that whatever necessarily includes a Kind is itself a Kind.

Possibility

We can formalize Boethius's various notions of repugnance as follows. The first notion, taken as a relation between terms rather than propositions, is expressed in definition 2.1.

Definition 2.1. '*a*' and '*b*' are repugnant$_1$ iff for some '*d*' EITHER $\Box\, a \mid d$ and $\Box\, b \rightarrow d$ OR $\Box\, a \mid d$ and $\Box\, d \rightarrow b$.

The difficulty, as signalled earlier, is that given the disjunction of these formulae we cannot conclude that $\Box\, a \mid b$. The second disjunct does indeed imply that result, by virtue of Axioms 1.2 and 1.8. But the first does not do so. So Boethius's first notion of repugnance is useless if we want repugnance to be a type of incompatibility.

The second notion is concerned with propositional rather than term logic, but appears to define repugnant terms simply as incompatibles. Its term logic is analogue is as follows.

Definition 2.2. '*a*' and '*b*' are repugnant$_2$ iff $\Box\, a \mid b$.

[43] In Ammonius (1998) 187, 10ff.

Turning to Boethius's third definition, 'a' and 'b' are repugnant provided that there are incompatible terms 'c' and 'd' of which the original pair of terms are consequences.

Definition 2.3. 'a' and 'b' are repugnant$_3$ iff for some 'c' and 'd' $\Box\ [a \leftarrow c \mid d \rightarrow b]$.

Again, we cannot conclude from this that $\Box\ a \mid b$. So we turn to the fourth definition, which makes the original terms antecedents rather than consequents of the interpolated incompatibles.

Definition 2.4. 'a' and 'b' are repugnant$_4$ iff for some 'c and 'd' $\Box\ [a \rightarrow c \mid d \leftarrow b]$.

And here again we have a definition that ensures that 'a' and 'b' are themselves incompatible. This can be shown using Axioms 1.2 and 1.8. The only trouble with this is that the result is trivial. *Every* incompatible pair will stand in the relation of repugnance by this account, because in every case it is possible to interpolate an incompatible pair, even if it has to be the original pair. The conclusion we must draw is that we might as well settle for the simple notion of incompatibility, rather than attempt to construct a notion in the way that Boethius attempted.

Statements of impossibility *cum determinatione* can be used to express Aristotelian impossibilities – such as the impossibility of chaff's burning *while it is at the bottom of the sea* or the impossibility of an animal's future seeing *once it has gone blind.* Given that the chaff is at the bottom of the sea, there is a property that it has which is incompatible with its burning. Given that the animal has gone blind, there is a property that it has which is incompatible with its seeing in the future. Formally,

Definition 2.5. 'a' is impossible for x iff for some 'd', $x \rightarrow d$ and $\Box\ d \mid a$.

Assuming that we may substitute '*not-a*' for 'd', this definition implies that if x is *not-a*, then 'a' is impossible for x, because '*not-a*' is incompatible with 'a'. Equally, assuming that we may substitute 'will always be *non-a*' for 'd', the definition implies that 'will at some time be a' is impossible for x, because 'will at some time be a' is incompatible with 'will always be *not-a*'. Combining these two results and contraposing, it follows that the possible for x is whatever x is or will at some time be. So, contrary to our intention (which was to analyse Aristotelian possibility) we have come up with a version of Diodorean possisbility. What has happened here is that a necessary condition for Aristotelian impossibility has been treated as if it were a sufficient condition. The condition in question is both necessary and sufficient for Diodorean impossibility. Thus we are left with the problem of finding a sufficient condition for Aristotelian impossibility.

The Aristotelian concept may be captured by amending Definition 2.5 so that we are prohibited from substituting for '*d*' terms that relate to the future.

Definition 2.6. '*a*' is impossible for *x* iff for some '*d*' such that '*d*' is not about the future, $x \rightarrow d$ and $\Box\, d \mid a$.

According to both Definition 2.5 and Definition 2.6, if Socrates is sitting, it is impossible that he now be standing, because his now standing is incompatible with his current attributes; but it is possible that he will be standing, because his standing in the future is so compatible. Equally, if Socrates is blind, both definitions imply that it is impossible for him to see, assuming that 'will be seeing' is incompatible with 'is blind'. In order to see more clearly the difference between the two definitions, it may be useful to consider Aristotle's example of a cloak that may either be cut up or wear out. Its being cut up is assumed to be incompatible with its wearing out. Suppose it will be wear out. Is it then possible for it to be cut up? According to Definition 2.5 this is not possible, but according to Definition 2.6 it is. We get the Diodorean answer if we allow 'will wear out' as something that *is true* of the cloak, because this predicate is incompatible with 'will be cut up', and so it follows from the definition that it is impossible for the cloak to be cut up. But there is something troubling about this answer in so far as it takes a predicate describing a future state of the cloak to be something that *is now true* of it. By eliminating this troubling feature we can get the Aristotelian answer using Definition 2.6. Given that the predicates now true of the cloak cannot describe the cloak's future states it seems that no predicate now true of it is incompatible with the supposition that it will be cut up, and so its being cut up is possible even if it will in fact wear out. So Definition 2.6, with this stipulation about what may be substituted for '*b*', allows for alternative future possibilities. Under the same stipulation it also allows that possibilities attaching to a species may be closed off for an individual member of that species, provided that they are incompatible with *something* true of the individual. The definition also accommodates Aristotle's dictum that what is, when it is, necessarily is. If Socrates is seated then something true of Socrates is incompatible with his not being seated, and so it is impossible that he is not seated (even though it is perfectly possible that he *will* not be seated). By this means we can recognize Boethius's class of necessity-propositions that are equivalent to assertorics. 'Socrates is seated' is equivalent to 'It is necessary that Socrates is seated'. We shall revisit the question of formalizing Philo's concept of possibility when we come to discuss Abélard.

Chapter 3

Abélard

Pierre Abélard (1079-1142) developed a modal logic that is wide-ranging and highly original. We get a glimpse of this level of originality in a remarkable passage from John of Salisbury's *Metalogicon*. John recollects something he had heard Abélard say about the art of modal logic:

> ...the Peripatetic of Pallet made the observation, which I believe was correct, that it would be easy for one of our own contemporaries to compose a book about this art, which would be at least the equal of any of those written by the ancients....[1]

Abélard was not an unduly humble man. If he thought that one of his contemporaries could equal the ancients in modal logic, he must have thought the same about himself. As we are about to see, this self-assessment is borne out by the contents of his major logical treatise, the *Dialectica*, and of the introductory essay to his commentary on chapters 12 and 13 of the *De Interpretatione*.[2]

---

[1] John of Salisbury (1962) III, pp.166-167.

[2] Abélard cannot have seen the relevant chapters of the *Prior Analytics*. On the other hand he knew that Aristotle had developed a modal syllogistic in that book. In both the *Dialectica* and the *Logica* Abélard makes an allusion to Aristotle's *Analytics*. De Rijk specifies 32a16, while Minio-Paluello refers to 30a15-40a3 – neither of which clarifies Abélard's cryptic allusion. If Abélard's 'possibile' is taken in the sense of one-sided possibility then the syllogisms in question are not asserted in the *Prior Analytics*, though they are equivalent to Aristotelian syllogisms. If it is taken in the sense of two-sided possibility then the syllogisms are asserted in the abstract and without concrete terms at A15, 33b33-36, A19, 38a16-18 and A21, 39b16-22. (These theorems are summarized without the use of variables in the opening sentences of A15, A19 and A21.) What he had to do, in effect, was to work out for himself what a modal syllogistic would have to be like, given some simple first principles. The logic that he constructed is not the same as Aristotle's, and his development of it is far from being complete, but in many respects it is superior to the modal syllogistic of the *Prior Analytics*. He might not have been so original had he possessed a fuller acquaintance with Aristotle's *Prior Analytics*.

Syntax

Abélard distinguishes between the syntactic features of a proposition *secundum constructionem* and *in sensu*. He asks two principal questions about modal propositions: What role does modality play in their syntactic structure, and. How are they derived from the corresponding assertorics? With regard to the first of these questions, Abélard's view is that modal words qualify neither the subject nor the predicate nor the whole proposition, but the relation between predicate and subject. On the second question, his view is that modal propositions are in sense *de rebus* – i.e. they share their subjects with the simple propositions from which they derive, rather than having the sense of those simple propositions as their subjects.[3] This is what it means to be *de rebus*. The doctrine that modal propositions are *de rebus* I will call actualism, since it claims that modal propositions are about the same actual things that are the subjects of assertoric propositions. Actualism, as we shall see, is not necessarily an anti-metaphysical doctrine.

What, then, of the *predicates* of modal propositions? What is the relation between the predicate of an unquantified modal proposition and that of the corresponding simple proposition? Here, Abélard's stated doctrine does not fully accord with his practice. The stated doctrine is that both terms of a modal proposition are in sense identical with the terms of the parent assertoric:

> In sense, modals have the same terms as simples. Indeed, when one says 'It is necessary for Socrates to run', this, which is properly modal in sense, has the same predicate and subject as the assertoric 'Socrates runs' from which it derives, because 'Socrates' is subject and 'runs' is predicate, and 'necessary' is in sense a determinant of the predicate (just as in its adverbial equivalent 'Socrates runs necessarily').[4]

This doctrine implies that the predicate of a modal proposition does not include the mode; however, Abélard speaks at times as if the predicate does include the mode:

> 'Socrates' is in sense the subject and 'runs necessarily' or 'can run' in sense are predicated.[5]

---

[3] Abélard (1956) 198:8-9. These words occur at the climax of a long polemic, 195:11-198:11, against 'his master's' *de sensu* analysis, according to which modals are about the sense of the corresponding simples.

[4] Abélard (1956) 191:26-27; Abélard (1958) §11.

[5] Abélard (1958) §12.

In considering quantified possibility-propositions, Abélard continues to distinguish between the subject and predicate according to the construction and the subject and predicate in sense.[6] His stated doctrine is that, in sense, all modal propositions – quantified or unquantified – have the same subject and predicate as the non-modal propositions from which they derive. In believing this, he has an eye to his paradigms of modal propositions. In those paradigms – for example, 'Socrates argues well' – a modal proposition is generated from a non-modal one by the insertion of a mode ('well') indicating how the verb coheres with the subject.[7] In these paradigm cases, modal and non-modal propositions share both subject and predicate. Abélard is indeed willing to count among modal propositions some that depart from his paradigms in various respects – for example, in containing nominal rather than adverbial modes ('For Socrates to run is possible' as well as 'Socrates possibly runs') – but in all cases he insists on the sharing of both terms between modals and their assertoric originals.[8] This doctrine is in agreement with Aristotle's practice in his modal syllogistic.

Abélard's stated doctrine clearly differentiates between a modal proposition's subject and predicate in sense, and according to the construction. It also appears to privilege the in-sense analysis over the constructional analysis. What is not clear is what such a privileging would amount to. Does Abélard think that the syntax *in sensu* is the real logical syntax of modal propositions while the syntax *secundum constructionem* is logically misleading? We will come back to this question.

Let us turn now to Abélard's well-known *de rebus/de sensu* distinction. He introduces the distinction by reference to Aristotle's analysis, in the *Sophistical Refutations*, of the sentence 'It is possible for the standing to sit'. Aristotle says this sentence is ambiguous between a compound and a divided sense; and Abélard identifies this distinction as the *de sensu/de rebus* distinction.

> ...as if we were to say 'It is possible for those standing to sit while remaining standing' i.e. 'to sit and stand simultaneously' (as if we were to say 'It is possible for things to happen as the proposition "Those standing are sitting" states').[9]

[6] Abélard (1956) 193:3-18, Abélard (1958) §11-14.
[7] Abélard (1958) §2, Abélard (1956) 191:7.
[8] Abélard (1958) §7, Abélard (1956) 191:9-26.
[9] Abélard (1958) §18.

Understood *de sensu* the proposition is not strictly speaking modal, since it is contrary to Abélard's stated doctrine of term-sharing between modals and their assertoric originals.[10] It is remarkable, then, that he goes on to maintain that the subject of a *de sensu* proposition – far from being a propositional sense – is the same as the subject of the corresponding simple proposition.

> It is to be noted also that, when we say that propositions are taken *de sensu*, it is not that we have a thought [*intellectus*] about the sense of a proposition, but [we think] only of the things, just as we do in that [proposition] which is *de rebus*.[11]

This change of mind seems to indicate a desire on Abélard's part to assimilate *de sensu* propositions to his preferred *de rebus* model, and a reluctance on his part to acknowledge them as exhibiting a genuinely distinct form. It seems, in reality, that the *de sensu/de rebus* distinction has a different *fundamentum divisionis* from the compound/divided distinction, and that in principle we can distinguish the four cases displayed in Table 3.1.

Table 3.1 *De sensu/de rebus*, compound/divided

De sensu compound:	'It is possible that: *b* is *a*-while-*b*'
De sensu divided:	'It is possible that: *b* is *a*'
De rebus compound:	'*b* can be *a*-while-*b*'
De rebus divided:	'*b* can be *a*'

While the first two of these express a possibility-that, the last two express a possibility-for.[12]

Abélard states that the compound *de rebus* proposition implies the divided one, but that the reverse implication does not hold.[13] Now, it is true that the two *de sensu* propositions are equivalent to one another: the possibility that *b* is *a* is the same as the possibility that *b* is *a*-while-*b*. But it is not true that either *de sensu* proposition is equivalent to either of the *de rebus* ones: the reason is that the latter imply the existence of a *b*, while the former do not. Thus there are three logically distinct propositions:

---

[10] Abélard (1958) §20. Abélard (1956) 195:21-26. Compare the discussion in Knuuttila (1993) pp.85-86.

[11] Abélard (1958) §46.

[12] The distinction is Marenbon's (2000) X, pp.507-599.

[13] Abélard (1958) §47, §49.

- *De sensu* ('The sense of the proposition "a *b* is *a*" is possible') (equivalent to 'It is possible for a *b* to be *a* while remaining *b*')
- *De rebus,* on the compounded reading ('Some *b* is possibly *a* and *b*')
- *De rebus,* on the divided reading ('Some *b* is possibly *a*').

Of these, the compound *de rebus* proposition implies both the *de sensu* and the divided *de rebus* propositions, but the implications are not reciprocated, nor is there an implication either way between *de sensu* propositions and divided *de rebus* ones.

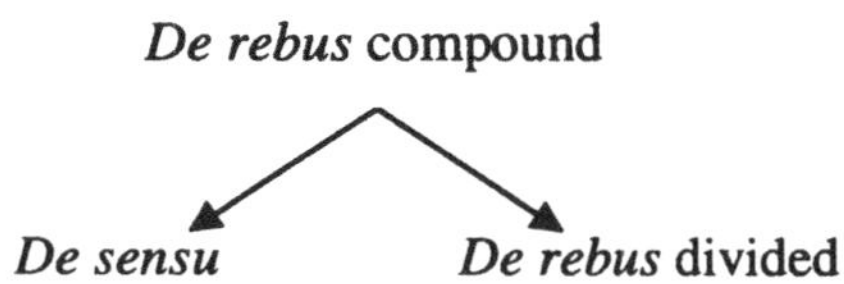

Figure 3.1 Types of *de rebus* proposition

Abélard however maintains that there are only two senses – *de sensu* and *de rebus* – and that the former implies the latter. He considers an objection to this view, citing the proposition 'It is possible for all substances to be spirits', which might be taken to be true *de sensu* but false *de rebus.* His response is to say that since the *de sensu* proposition is equivalent to the compound, and since the latter implies the divided proposition, the *de sensu* proposition implies the *de rebus*, and so the proposition 'It is possible for all substances to be spirits' is false, not true.[14] The trouble is that Abélard's own counter-example convincingly shows the opposite to be the case. The proposition asserting the possibility of a situation in which all substances are spirits is a different proposition from that asserting that all currently existing substances could be spirits (whether the latter proposition is taken in the compound or the divided sense).

[14] Abélard (1958) §48-49.

Restricted modals

Abélard distinguishes both *de rebus* and *de sensu* modals from another type, those that Boethius had called modals *cum determinatione* – namely modals that include a phrase beginning with the word '*dum*' ('so long as').[15] Abélard distinguishes two species. The restriction is *intrasumpta* when the predicate of the simple proposition from which the modal is derived is repeated in the modal. 'It is possible for Socrates to be a man, so long as he is a man' is of this type. The restriction is *extrasumpta* when the predicate of the modal is other than that of the simple proposition from which the modal is derived. 'It is possible for Socrates to be a man, so long as he lives' is of this type. Abélard states that for the truth of any of these modals it is required that the restricting clause (the *dum*-clause) be true, and this fact distinguishes restricted modals from modals *de sensu*: 'It is possible for Socrates to be a man, so long as he is a man' implies that Socrates is a man, whereas 'It is possible for Socrates to be man while remaining a man' does not have this implication.

Abélard states[16] – as against Boethius's view – that restricted modals are not equivalent to their parent assertorics, whether their restrictions are *extrasumpta* or *intrasumpta*. He argues that if both 'It is possible for Socrates to read so long as he reads' and 'It is necessary for Socrates to read so long as he reads' are equivalent to 'Socrates reads', then they are equivalent to each other. But these are not equivalent since possibility does not imply necessity. Abélard concedes that one of these restricted modals is never true without the other being true, but he draws a distinction between *comitatio* and *inferentia*, stating that in the present instance we have *comitatio* but not *inferentia*. We shall return to this distinction.[17]

Semantics

Abélard intermittently uses a semantics that relies on the notions of what a thing's nature allows, and what is repugnant to a thing's nature. For example, the *Dialectica* expounds the proposition 'It is possible for Socrates to be a bishop' as stating that being a bishop is not repugnant to

[15] Abélard (1956) 206:14ff.

[16] Abélard (1956) 207:19ff.

[17] The logic of restricted modals is developed at considerable length by Abélard, but since it does not bear on his modal syllogistic it will not be discussed here.

Socrates' nature.[18] In Abélard's thinking a predicate '*a*' is impossible for a subject *b* if *b* has a nature '*n*' that is repugnant to '*a*'. Thus, Abélard's notion of possibility rests on two fundamental concepts.

Repugnancy. We might understand this as the relation that Abélard, following Boethius, defines as holding between *repugnantia*. Abélard's definition of this relation follows the fourth of Boethius's definitions, noted in Chapter 2.[19] Given a pair of contraries such as being awake and being asleep, snoring, which is subordinate to being asleep, is repugnant to being awake. This definition of repugnancy is, however, not quite what is needed in the context of Abélardian possibility, for the following reason. It is possible for the standing not to be standing. However, there is no contrary of 'standing' to which 'not standing' is subordinate, since any contrary of 'standing' would be subordinate to 'not standing'. It is better, then, not to adopt the Boethian definition of repugnancy, but to work with a wider sense according to which the repugnancy between a given predicate and a thing's nature is simply a matter of the two being *incompatible*. This is a purely logical and a *de sensu* notion. Abélardian (*de re*) possibility therefore depends on a *de sensu* notion of incompatibility. Abélard, however, is silent about this dependence; and his silence perhaps indicates a failure to appreciate the importance of *de sensu* propositions even for his own *de rebus* modal theory.

A thing's nature. Abélard links the idea of a thing's nature with what he calls substantive predications.[20] The latter, he explains, occur where a species is predicated of its individuals, or a genus of its species.[21] A thing's nature, it seems, is the species or genus to which the thing belongs. In our formalization, a thing's nature is expressed by a *per se* term, and to that extent is a metaphysical notion. Talk of a thing's nature involves an implicit quantification, since an individual can belong to more than one species: for example, Socrates is by nature both a man and an animal. What is impossible for him is what is incompatible with *some* species to which the individual belongs; what is possible for him is what is compatible with *any* species to which he belongs. Thus we have the definition:

---

[18] Abélard (1956) 193:33-194:5. Abélard is here following Garlandus Compotista, who had advanced a notion of a thing's potentialities *extra actum* as 'that which it will never be, but to which its nature is not repugnant', and had given the example 'It is possible for Garlandus to be made a bishop'. Garlandus Compotista (1959) 83:35-84:1.

[19] Abélard (1956) 450:33-451:3. See p.38 above.

[20] Abélard (1956) 361:19-20.

[21] Abélard (1956) 425:15-16.

Definition 3.1. 'a' is possible for x ($x \rightarrow a^{\dagger}$) iff there is no *per se* term 'n' such that $x \rightarrow N$ and $\Box\, n \mid a$. This differs from Definition 2.5 in requiring that the intermediary between a subject and that which is impossible for it must be a term denoting a nature.

It is worth noting that the statement that Socrates is possibly a bishop can be represented as making a certain complex predication of Socrates – namely to predicate of him that he has a nature compatible with being a bishop. Abélard does not explicitly represent statements of possibility in this way; but we shall see that this manner of representation is quite important for his modal theories.

Why does Abélard think it is not repugnant to Socrates' nature that he be a bishop? Because there are other individuals, of the same species as Socrates, who actually are bishops. Abélard enunciates the general theorem that what actually happens in one member of a species is possible for any member of the species.[22]

Theorem 3.1 (Abélard's Thesis). If $N \leftarrow x \rightarrow a$ then if $y \rightarrow N$ then 'a' is possible for y. This is in fact a logical consequence of Abélard's definition of possibility (though he himself did not demonstrate this). Suppose that some man is a bishop. Then being a bishop is compatible with all species to which that man belongs. However, the species to which that man belongs are the same as the species to which Socrates belongs. Therefore, if what is possible for Socrates is defined as what is compatible with all species to which Socrates belongs, it follows that being a bishop is possible for Socrates.

It might be thought that for Abélard, possibility is always *dynamic* in the sense that what is possible for a subject is what that subject can *become*. However, not all Abélardian possibilities are dynamic in this sense. First of all, Abélard recognizes that the inference *ab esse ad posse* is valid: he accepted that what an individual *is*, has to be included among its possibilities.[23] Indeed, he is logically committed to recognizing this: if Socrates is a bishop then some man is a bishop, and if some man is a bishop then (by Abélard's Thesis) Socrates can be a bishop, therefore (from the first to the last) if Socrates is a bishop then Socrates can be a bishop. A second way in which Abélardian possibility may be non-dynamic arises in connection with irreversible changes. Abélard states that after someone has become blind it is possible for him or her to see.[24] This is not because the blind can come to see but because seeing is not repugnant to the species to

---

[22] Abélard (1956) 193:34-194:5. As Marenbon (2000) notes (X, p.600) this theorem states a sufficient, not a necessary, condition.

[23] See Abélard (1956) 204:20-21.

[24] Abélard (1958) §75; Abélard (1956) 385:11-14.

which the blind belong.[25] Abélard's definition of possibility does indeed imply that if sightedness is not repugnant to the nature of the human species then it is possible for *all* humans – even for those who can never *become* sighted – to see.

Given Alexander's classification of definitions of possibility (noted in the previous chapter), Abélard's account seems to be Philonian rather than Diodorean or Aristotelian. Given that chaff is by nature combustible, Abélard would have to say that any quantity of chaff, no matter how situated, can be burnt, so long as it remains chaff. Abélard's Thesis – that what is actual in one member of a (most specific) species is possible for all members of that species – does not hold for Diodorean possibility. What will be actual for one member of a species may not become actual for all. One horse, and only one, will win the race, and by Diodorus's lights winning was never possible for the ones that lost. Equally, Abélard's Thesis fails for Aristotelian possibility. According to Aristotle the possibilities for those who have become blind are not the same any more as those for sighted people. Sighted people actually see; but that fact does not establish that those who have become blind can see.

The Philonian notion, unlike the Diodorean and Aristotelian notions, is a notion of abstract, or species-wide, possibility. While this is a recognizable element in our ordinary understanding of possibility, it clearly is not rich enough to encompass all the judgements of possibility that we ordinarily want to make. If this were our only notion of possibility, we would not be able to make judgements of *concrete* possibility, such as the judgement that the chaff at the bottom of the ocean *cannot* burn *there*. The chaff cannot burn there because it has been prevented from burning by an external necessity. So the question arises whether Abélard has any means for making such concrete judgements. I believe that the answer to this question is affirmative. Abélard can invoke his distinction between compound and divided modals at this point. In the compound sense, it is false to say that chaff at the bottom of the ocean can burn-while-being-at-the-bottom-of-the-ocean, because burning-while-being-at-the-bottom-of-the-ocean is incompatible with the nature of chaff. Indeed, since burning-while-at-the-bottom-of-the-ocean is an impossibility, it is incompatible with everything. In the divided sense, however, it is true to say that chaff at the bottom of the ocean can burn, because burning is not incompatible with the nature of chaff at the bottom of the ocean. Chaff at the bottom of the ocean has precisely the same nature as chaff on dry land.

---

[25] Marenbon (2000) X, n.18. Compare Abélard (1927) 273:39-274:4.

Equipollence and Opposition

Abélard correctly describes the logical relations and equipollences among modal propositions with singular subjects. Those logical relations may be summarized in a diagram such as Figure 3.2 (where 'L' is short for 'it is necessary', 'M' for 'it is possible', '~L' for 'it is not necessary' and '~M' for it is not possible'). In this diagram, implications are represented by arrows, and relations of contradiction are represented by dotted lines. The diagram departs from what has become the standard way of drawing such diagrams, with contradictories diagonally opposite, and with contraries explicitly represented,[26] but is closer to what Aristotle appears to have had in mind in the *De Interpretatione*, where contradictories are placed side by side;[27] it is also more elegant than the traditional square of opposition, in omitting mention of contraries (which are definable in terms of implication and contradiction, a contrary of *p* being any proposition that implies ~*p*).

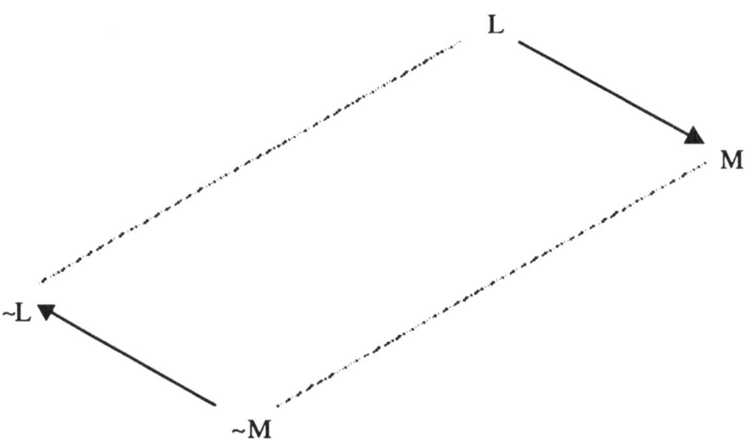

Figure 3.2 Modal square of opposition

Abélard recognizes the equipollences 'L = ~M~', 'L~ = ~M', 'M = ~L~' and M~ = ~L'.[28] He takes possibility-propositions as the primary case, with impossibility- and necessity-propositions regarded as secondary. Possibility-propositions affirm or deny either a possibility to-be or a possibility not-to-be. In the first case we have what Abélard calls a proposition *de esse*, in the second case a proposition *de non esse*. It is

---

[26] See for example Thom (1981) §19.

[27] Aristotle (1963) 19b19ff, ch.13, 22a14ff.

[28] Abélard (1956) 198:12-20, 198:32-199:24; Abélard (1958) §35-38.

striking that in using the *de esse/de non esse* distinction here, Abélard is taking singular modals according to their construction, not 'in sense'. In doing so, however, he is not misleading us about the logic of these propositions, and that fact casts doubt on any suggestion that the *secundum constructionem* analysis is logically misleading. Singular modals, it seems, can equally well be analysed according to their construction or in sense. They genuinely possess the logical features displayed in either analysis. These features can be described in modern language as follows. The question whether a singular possibility-proposition is negative or affirmative is the question whether there is an external negation; the question whether it is *de non esse* or *de esse* is the question whether the possibility-sign is immediately followed by a negation. The proposition genuinely possesses both features. At the same time, the negative/affirmative feature has a certain priority, in the sense that it has the wider scope.

Corresponding to Definition 3.1, and given that 'necessary to be' is equipollent to 'not possible not to be', we can define what is necessary for a given individual as that whose negation is incompatible with the thing's nature. Equivalently:

Definition 3.2. 'a' is necessary for x ($x \rightarrow a^{*}$) iff there is a *per se* term 'n' such that $x \rightarrow N$ and $\Box\, n \rightarrow a$.

Abélard takes quantified modal propositions, understood *de rebus*, to possess quantity and quality in the same way as assertorics.[29] For example, he takes the proposition 'For some man it is not possible to be white' to be a particular negative whose meaning is 'Some man's nature is repugnant to "white"'.[30] This is equivalent to '"White" is impossible for some man', a particular negative proposition with subject 'man' and predicate 'possibly white'. Following this line of analysis, and making use of definitions 3.1 and 3.2, we state truth-conditions for eight forms of quantified possibility-proposition in Definitions 3.3 to 3.10.

[**Definition 3.3 (La).** 'Every b is necessarily a': $b \rightarrow a^{*}$.][31]

Definition 3.4 (Le). 'No b is possibly a': $b \mid a^{\dagger}$.[32]

[**Definition 3.5 (Li).** 'Some b is necessarily a': for some 'd', $b \leftarrow d \rightarrow a^{*}$.]

Definition 3.6 (Lo). 'Some b is not possibly a': for some 'd', $b \leftarrow d \mid a^{\dagger}$.[33]

---

[29] Abélard (1956) 192:31-193:2; Abélard (1958) §14.

[30] Abélard (1956) 200:22-25.

[31] Analyses of the bracketed modal forms are not given explicitly in either *Dialectica* or *Logica*.

[32] Abélard (1956) 200:28-30: ' "For no man is it possible to be white", i.e., "Every man's nature is repugnant to 'white'".' Compare Abélard (1958) §44.

Definition 3.7 (Ma). 'Every *b* is possibly *a*': $b \rightarrow a^{\dagger}$.[34]
Definition 3.8 (Me). 'No *b* is necessarily *a*': $b \mid a^{*}$.[35]
Definition 3.9 (Mi). 'Some *b* is possibly *a*': for some '*d*', $b \leftarrow d \rightarrow a^{\dagger}$.[36]
[**Definition 3.10 (Mo).** 'Some *b* is not necessarily *a*': for some '*d*', $b \leftarrow d \mid a^{*}$.]

These definitions are purely logical in their content. They are all that we need in order to work out Abélard's modal logic, though (as we have seen) they are underpinned by his metaphysical definition of possibility (Definition 3.1).

Equipollence of quantified modals

The equipollences among quantified modal propositions are shown in Tables 3.2 and 3.3.

When it comes to our eight quantified possibility-propositions, Abélard again invokes the *de esse/de non esse* distinction, and with it the *secundum constructionem* analysis. He distinguishes eight propositional forms, according as they are universal or particular, affirmative or negative, *de esse* or *de non esse*.[37] He says that it is easy to work out the relations of contradiction and implication.[38] And so it is, given his semantics for *de rebus* modals. For example, the contradictory of *Ma* is *Lo* because if it is the case that no *b* has a nature with which '*a*' is incompatible then it is not the case that some *b* has a nature with which '*a*' is incompatible, and *vice versa*. There is no reason to doubt that Abélard knew the correct answers to questions about the logical relation between any two of his eight modal forms. His semantics support the construction of two modal squares of opposition (see Figures 3.3 and 3.4), one concerning modal propositions *de esse*, the other concerning modals *de non esse*. There is a doubt, however, about whether he knew how to construct squares of opposition for these forms.

---

[33] Abélard (1956) 200:23-25; Abélard (1958) §44. 'When we say "It is impossible for some thing to be a man", the proposition is true when taken *de re*, because "man" is repugnant to a certain thing's nature.'
[34] Abélard (1956) 200:26-28. Compare Abélard (1958) §45: 'This is how they are to be expounded *de rebus* for individual modes: "It is possible for every thing to be a man", i.e., … "Each thing's nature allows it to be a man".'
[35] Abélard (1956) 201:27-28.
[36] Abélard (1956) 200:30-31.
[37] Abélard (1958) §40. Compare Abélard (1956) 199:25-201:24.
[38] Abélard (1956) 202:27-29; Abélard (1958) §41.

In fact however, Abélard did not list the logical relations in these two squares of opposition.[39] What is worse, in the *Logica* he stated that the logical relations among the four universal forms (Ma, Le, Me, La) are the same as those among the four particulars (Mi, Lo, Mo, Li), and that these are both the same as the logical relations among the four unquantified forms.[40] This, as Knuuttila says, is 'totally wrong',[41] and it suggests that while Abélard knew the oppositions among universal and particular modals individually, he was unable correctly to collect them into a square of opposition.

Table 3.2 *De esse* modals

Ma	*Le*	*Mi*	*Lo*
'For every man it is possible to run'	'For no man is it possible to run'	'For some man it is possible to run'	'For some man it is not possible to run'
'For no man is it impossible to run''	'For every man it is impossible to run'	'For some man it is not impossible to run'	'For some man it is impossible to run'
'For no man is it necessary not to run'	'For every man it is necessary not to run'	'For some man it is not necessary not to run'	'For some man it is necessary not to run'

Table 3.3 *De non esse* modals

Me	*La*	*Mo*	*Li*
'For every man it is possible not to run'	'For no man is it possible not to run'	'For some man it is possible not to run'	'For some man it is not possible not to run'
'For no man is it impossible not to run''	'For every man it is impossible not to run'	'For some man it is not impossible not to run'	'For some man it is impossible not to run'
'For no man is it necessary to run'	'For every man it is necessary to run'	'For some man it is not necessary to run'	'For some man it is necessary to run'

[39] See Jacobi (1980) pp.139-140.
[40] Abélard (1958) §41.
[41] Knuuttila (1993) p.88.

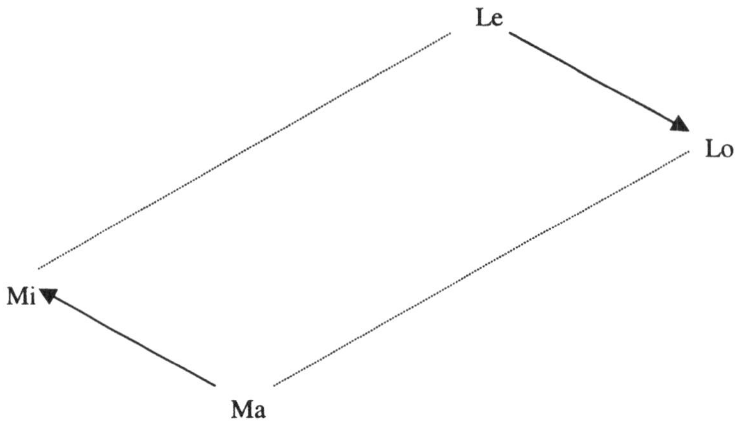

Figure 3.3 Square of opposition for *de esse* modals

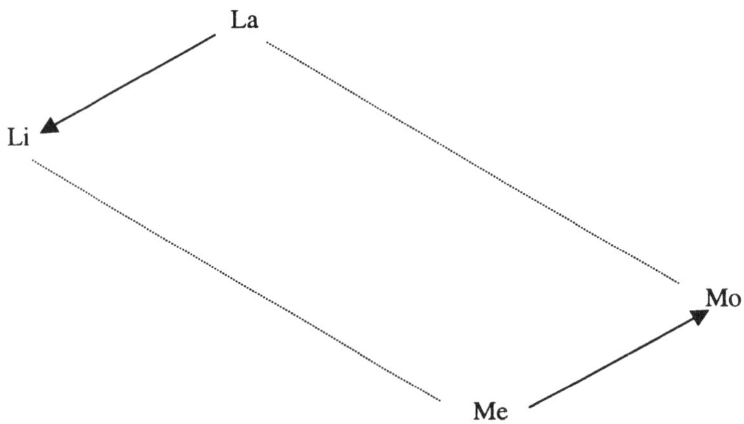

Figure 3.4 Square of opposition for *de non esse* modals

The quality and quantity of quantified possibility-propositions are genuine logical features of them, as is the feature by virtue of which they are *de esse* or *de non esse*. The latter is the question whether the predicate is negative; and, as with singular possibility-propositions, this question is secondary to the question of quality and quantity, in the sense that the feature it concerns is a feature of narrow, not wide, scope.

Subordination and Conversion

Abélard knew the standard laws of modal subordination (from 'It is necessary' to 'It is so' and from 'It is so' to 'It is possible').[42] It follows from these laws that a *de rebus* necessity-proposition implies the corresponding assertoric proposition, and that this implies the corresponding *de rebus* possibility-proposition.

Abélard deals with the conversion of modal propositions,[43] arguing against those who favour a *de sensu* analysis of modals. On their analysis, says Abélard, modal propositions should obey the laws of simple conversion in the same way as assertorics do, since if an assertoric is possible or impossible or necessary, then so is its equivalent converse. Yet these people deny various modal conversions, for example they say that while it is possible for every non-stone to be a non-man (since every non-man is already a non-man, and every man will become such), it is not possible for every man to be a stone. Abélard points out rightly that, on their *de sensu* analysis, they should say that the antecedent as well as the consequent here is false: the proposition 'Every non-stone is a non-man' is *not* possible. The point is well-taken; but it is purely *ad hominem*, and reveals nothing of Abélard's own views on modal conversion.

For those we must turn to the *Logica* where Abélard argues that conversion holds in *de rebus* modals no less than in simple propositions, provided we pay careful attention to their sense. The context makes it plain that modal propositions convert in the same ways as non-modals if and only if we take the mode to be part of the predicate, and that if we do this then the converse is not a modal proposition.[44] Abélard gives examples of several of these inferences: 'For every man it is possible to run' converts to 'Something for which it is possible to run is a man', 'For no body is it necessary to be a man' converts to 'Nothing for which it is necessary to be a man is a body'.[45]

The converting proposition here shares its predicate with its original assertoric only in the loose sense that its predicate *includes* that of the assertoric. If we were to take the predicate of 'For every man it is possible to live' as 'live' (as required by Abélard's doctrine of term-sharing between modals and their assertoric originals), the question of contraposition would then be whether the premise implies any proposition about the non-living. Abélard does not explicitly consider this question, but it is easy to see that

---

[42] Abélard (1956) 204:20-21.
[43] Abélard (1956) 195:15-16, 195:27-196:28; Abélard (1958) §16-17.
[44] Abélard (1958) §16.
[45] Abélard (1958) §16.

the answer must be negative. The stated premise does not imply any proposition about the non-living, as distinct from the possibly non-living. Similarly, it is evident from the semantics of Abélardian *de rebus* modals that the usual laws of conversion do not hold, if we take the predicate as excluding the mode. Given that for Abélard in genuinely modal propositions the predicate excludes the mode, we might describe his position as stating that modal propositions, *qua* modal, do not convert. They convert only by virtue of the fact that, when their predicates are seen as including a mode, they instantiate non-modal forms.

Modal syllogisms

Both *Dialectica* and *Logica* introduce a brief discussion of 'mixed' modal syllogisms by noting the validity of a concrete syllogism in Barbara MXM. The example in the *Logica* is 'For every animal it is possible to move, but every man is an animal, so for every man it is possible to move'.[46] Abélard does not state the syllogism formally, nor does he provide a proof; but both its form and its proof are easy to discern, relative to Abélard's modal semantics.

Theorem 3.2. Barbara MXM is valid. Proof: if $b \rightarrow a^{\dagger}$ and $c \rightarrow b$ then we have the chain '$c \rightarrow b \rightarrow a^{\dagger}$', which simplifies by Axiom 1.3 to '$c \rightarrow a^{\dagger}$'.

Using the principles of indirect reduction, we can generate the following corollaries, though they are not explicitly noted by Abélard.

[**Theorem 3.2a.** Baroco MLX is valid.]

[**Theorem 3.2b.** Bocardo LXL is valid.]

Abélard also gives an example of a Cesare LMX: 'For no stone is it possible to live, but for every man it is possible to live, so no man is a stone'.[47] Thus we have the theorem:

Theorem 3.3. Cesare LMX is valid. Proof: if $b \mid a^{\dagger}$ and $c \rightarrow a^{\dagger}$ then we have the chain '$c \rightarrow a^{\dagger} \mid b$', which simplifies by Axiom 1.4 to '$c \mid b$'.

This generates the following pair of corollaries:

[**Theorem 3.3a.** Ferio LXL is valid.]

[**Theorem 3.3b.** Datisi MXM is valid.]

He also gives an example of a Darapti MXM, but we will leave that to one side, in accordance with our policy of dealing only with fundamental

---

[46] Abélard (1958) §15.

[47] Abélard (1958) §15.

syllogisms.[48] These, however, are not the only modal syllogisms that Abélard mentions. The *Dialectica* adds that the first-mentioned syllogism works with necessity instead of possibility, i.e. Barbara LXL.[49]

Theorem 3.4. Barbara LXL is valid. Proof: if $b \rightarrow a^*$ and $c \rightarrow b$ then we have the chain '$c \rightarrow b \rightarrow a^*$', which simplifies by Axiom 1.3 to '$c \rightarrow a^*$'.

What would Abélard have said to the purported counter-examples that Alexander brought against this mood?[50] On the one hand, he would have accepted the truth *de rebus* of the conclusion 'Everything moving is necessarily an animal', given the premises 'Every man is necessarily an animal' and 'Everything moving is a man'; and on the other hand, he would have denied the truth *de rebus* of the major premise 'Everything walking is necessarily moving'. The latter is true merely *de sensu*.

Barbara LXL generates two more syllogisms:

[**Theorem 3.4a.** Baroco LMX is valid.]

[**Theorem 3.4b.** Bocardo MXM is valid.]

In another passage in the *Logica* he articulates a Celarent MXM: 'For no body is it necessary to be white, but every swan is a body, so for no swan is it necessary to be white'.[51]

Theorem 3.5. Celarent MXM is valid. Proof: if $b \mid a^*$ and $c \rightarrow b$ then we have the chain '$c \rightarrow b \mid a^*$', which simplifies by Axiom 1.4 to '$c \mid a^*$'.

This gives two more syllogisms:

[**Theorem 3.5a.** Festino MLX is valid.]

[**Theorem 3.5b.** Disamis LXL is valid.]

The *Logica* includes an incomplete statement of a Celarent LXL. 'For every thing it is impossible to be a man' is said to imply 'For every man it is impossible to be a man. This becomes a Celarent LXL if we add the obvious assumption that every man is a thing.[52]

Theorem 3.6. Celarent LXL is valid. Proof: if $b \mid a^{\dagger}$ and $c \rightarrow b$ then we have the chain '$c \rightarrow b \mid a^{\dagger}$', which simplifies by Axiom 1.4 to '$c \mid a^{\dagger}$'.

Hence by indirect reduction we have:

[**Theorem 3.6a.** Festino LMX is valid.]

[**Theorem 3.6b.** Disamis MXM is valid.]

Now, given the validity of Barbara LXL, Celarent LXL, and Ferio LXL, we would expect Darii LXL also to be valid. Indeed, if the first three are

---

48 Abélard (1956) 245:23-246:2, Abélard (1958) §15. The *Dialectica* examples use the terms 'just', 'good', 'bad', 'virtue', which are familiar terms in Boethius's examples but do not figure in Aristotle's modal syllogistic.

49 Abélard (1956) 245:29.

50 See above p.28.

51 Abélard (1958) §11.

52 Abélard (1958) §44.

just non-modal perfect syllogisms with the mode of necessity or possibility attached to the major term (as they are), then Darii LXL will be valid for the same reason.

[Theorem 3.7. Darii LXL is valid.] Proof: if $b \rightarrow a^*$ and for some 'd', $c \leftarrow \underline{d} \rightarrow b$, then we have the chain '$c \leftarrow \underline{d} \rightarrow b \rightarrow a^*$', which simplifies by Axiom 1.3 to '$c \leftarrow \underline{d} \rightarrow a^*$'.

And from this we infer:

[Theorem 3.7a. Camestres LMX is valid.]

[Theorem 3.7b. Ferison MXM is valid.]

By the same token, given Barbara and Celarent MXM we would expect Darii MXM.

[Theorem 3.8. Darii MXM is valid.] Proof: if $b \rightarrow a^{\dagger}$ and for some 'd', $c \leftarrow \underline{d} \rightarrow b$, then we have the chain '$c \leftarrow \underline{d} \rightarrow b \rightarrow a^{\dagger}$', which simplifies by Axiom 1.3 to '$c \leftarrow \underline{d} \rightarrow a^{\dagger}$'.

Whence:

[Theorem 3.8a. Camestres MLX is valid.]

[Theorem 3.8b. Ferison LXL is valid.]

Analogous reasoning applies to Ferio MXM:

[Theorem 3.9. Ferio MXM is valid.] Proof: if $b \mid a^*$ and for some 'd', $c \leftarrow \underline{d} \rightarrow b$, then we have the chain '$c \leftarrow \underline{d} \rightarrow b \mid a^*$', which simplifies by Axiom 1.4 to '$c \leftarrow \underline{d} \mid a^*$'. It is striking that in this and all our other proofs of Abélard's theorems, we have used nothing more than simplification by Axioms 1.3 and 1.4. Thus, from the standpoint of Abélard's semantics, all these syllogisms (not just the first-figure ones) are perfect.

So:

[Theorem 3.9a. Cesare MLX is valid.]

[Theorem 3.9b. Datisi LXL is valid.]

Thus we have 24 basic valid mixed syllogisms. They are shown in Table 3.4.

Table 3.4 Abélard's system: syllogistic L/M/X combinations

Figure	*Combinations*
1	LXL, MXM
2	LMX, MLX
3	LXL, MXM

All the syllogisms in the table are valid on the Abélardian semantics sketched earlier. Because of this, every one of them can be seen as instantiating a valid non-modal syllogistic form, provided that in their premises and conclusions the modes are included within the predicates. For instance, Barbara MXM can be represented as a substitution in non-modal Barbara, where the major term is 'has a nature compatible with "*a*"'. Thus, in Abélard's syllogistic theory, just as in his theory of modal conversion, modal propositions are not treated specifically as modal; instead, their logic flows from the fact that they instantiate assertoric forms.

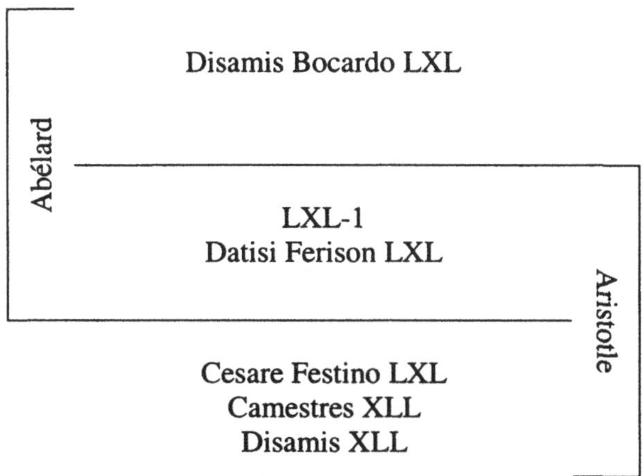

Figure 3.5 Abélard's L/X/M system and Aristotle's system

Aristotle's and Abélard's systems are compared in Figure 3.5. Valid for Abélard but not for Aristotle are Disamis and Bocardo LXL and their equivalents;[53] on the other hand, Aristotle accepts, and Abélard rejects, Cesare and Festino LXL, and Camestres and Disamis XLL – and their equivalents. It should be noted that the Aristotelian syllogisms Datisi and Ferison LXL, though valid for Abélard, cannot be proved valid by Aristotelian means.

In addition to the fundamental syllogisms, some inferences have strengthened premises or weakened conclusions, by virtue of modal subalternation or subordination. Subalternation is the inference from a universal to a particular proposition with the same modality, for example from 'All *b* are necessarily *a*' to 'Some *b* are necessarily *a*'. The use of

[53] Aristotle (1989) rejects Bocardo LXL at A11, 32a4-5, and Disamis LXL at A11, 31b27-33. His proposed counter-examples (if some animal cannot be a biped and every animal is moving it does not follow that something moving cannot be a biped, and if some animal must be a biped and every animal is awake it does not follow that something awake must be a biped) would not have troubled Abélard, who would have said that if the premises are true then so is the conclusion.

subalternation expands the set of syllogisms to include various forms of Darapti, Felapton, Barbari, Celaront and so on. (One of these – Darapti MXM – is explicitly accepted by Abélard.) This expansion of the basic stock of syllogisms is unproblematic for Abélard.

Inferences generated by modal subordination are another matter. It seems that Abélard would not accept such syllogisms, because (if we take the mode to be part of the predicate) there is no middle term. For instance by applying modal subordination to the minor premise of Barbara LXL we get Barbara LLL, where the subject of the major is '*b*' but the predicate of the minor is 'necessarily *b*'. This inference is, in the words of the *Logica*, 'in no figure'. The *Dialectica* considers whether such inferences are syllogistic, and concludes 'But indeed it seems to me that those that are made up solely from modals are not syllogisms.'[54] Syllogisms they may not be, but they must be valid inferences if modal subordination is valid (as Abélard believes). Abélard himself acknowledges as much, commenting 'though those that are made up exclusively from modals may not rightly be said to be syllogisms, still they hold maximum probability'.[55] It seems that in the case of these inferences we have *comitatio* but not *inferentia*. Applying our semantic analysis, we can prove the validity of Barbara LLL: if $b \rightarrow a^*$ and $c \rightarrow b^*$ then by interpolating Axiom 1.7 we may form the chain '$c \rightarrow b^* \rightarrow b \rightarrow a^*$', which by Axiom 1.3 simplifies to '$c \rightarrow a^*$'. Notice that in this proof, unlike our proofs of Abélard's L/X/M syllogisms, use is made of interpolation. In any case, wherever a syllogism in LXL is accepted by Abélard, the corresponding LLL syllogism is also valid (even if not syllogistically valid). This applies to all the syllogisms in the first and third figures. (In the second figure, however, there are no valid LLL syllogisms on Abélardian semantics.) And to each LLL syllogism there correspond two other syllogisms equivalent to it by indirect reduction, as shown in Table 3.5.

Table 3.5 Abélard's system: non-syllogistic L/M combinations

Figure	*Combinations*
1	LLL, MLM
2	LMM, MLM
3	LLL, MLM

[54] Abélard (1956) 246:23-24.
[55] Abélard (1956) 247:26-28.

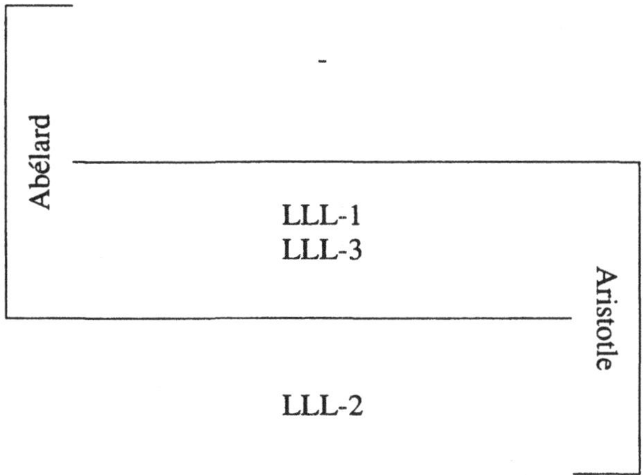

Figure 3.6 Abélard's L/M system and Aristotle's system

[Theorem 3.10. The LLL moods are valid in figure 1.] Barbara: if $b \rightarrow a^*$ and $c \rightarrow b^*$ then $c \rightarrow a^*$. Celarent: if $b \mid a^\dagger$ and $c \rightarrow b^*$ then $c \mid a^\dagger$. Darii: if $b \rightarrow a^*$ and for some '*d*', $c \leftarrow \underline{d} \rightarrow b^*$, then $c \leftarrow \underline{d} \rightarrow a^*$. Ferio: if $b \mid a^\dagger$ and for some '*d*', $c \leftarrow d \rightarrow b^*$, then $c \leftarrow d \mid a^\dagger$. In all cases, we can interpolate Axiom 1.7 ($b^* \rightarrow b$), and use Axiom 1.3 or 1.4.
[Theorem 3.10a. The LMM moods are valid in Figure 2.]
[Theorem 3.10b. The MLM moods are valid in Figure 3.]
[Theorem 3.11. The LLL moods are valid in Figure 3.] The proof will be as for Theorem 3.10.
[Theorem 3.11a. The MLM moods are valid in Figure 1.]
[Theorem 3.11b. The MLM moods are valid in Figure 2.]

The relations between Abélard's system of L/M moods and Aristotle's system are shown in Figure 3.6. (Only the LLL moods are listed, not their equivalents in LMM or MLM.)

Summary

Abélard has a clear conception of what a modal logic should look like. This conception rests on three basic ideas – a *de rebus* analysis of modal propositions, a definition of the possible as what is compatible with a thing's nature, and a distinction between *inferentia* and *comitatio*. Elaborating this conception, we can construct a system based on the mixed necessity/assertoric syllogisms in Figures 1 and 3. These are all valid by virtue of being substitution-instances of non-modal syllogisms. The pure necessity-syllogisms of those figures could be added as a derivative case (generated by modal subordination), but will have the status only of *comitatio* not *inferentia*.

The *de rebus* analysis is intuitively attractive because it validates the first-figure mixed necessity-syllogisms; however, it does not validate the laws of modal conversion or the modal syllogisms dependent on them, thus creating rather than solving a problem for interpreters of Aristotle. Another problem that it generates is the acceptance of the unAristotelian Disamis and Bocardo LXL. Abélard's distinction between *inferentia* and *comitatio* also creates a problem, in so far as it excludes the pure necessity-syllogisms. The distinction appears to be excessively exiguous in its requirements for syllogistic validity, by excluding the use of interpolation.

Chapter 4

Avicenna

As a modal logician, Avicenna (Ibn Sīna, d. 1037) shares a certain boldness with Abélard: neither man flinches at the prospect of opposing Aristotle. Avicenna hints at this boldness when, having just espoused a handful of unAristotelian doctrines, he adds 'the truth in this matter requires that about which we need not be diffident'.[1] Ironically, while the content of that remark is anti-Aristotelian, the style echoes Aristotle's statement of principle (invoked in the course of his critique of Plato), according to which 'piety requires us to honour truth above our friends'.[2] Avicenna's originality is evident in his unorthodox treatment of the semantics of assertoric propositions, in his systematization of the logic of ampliated modal propositions, and in his sketches towards a logic of those modal propositions that are qualified with phrases such as 'so long as'. In the present chapter, I shall be relying on the work of scholars such as Street who have worked with Avicenna's texts in the original languages. The works of Avicenna that we will discuss are his *Salvation* and *Pointers*.

Non-modal logic

Avicenna analyses universal assertoric propositions in an unconventional way, modalizing both subject and predicate. He says of a proposition with subject '*j*' that it is about 'every single thing described as *j*, be it in mental supposition or extra-mental existence, be it described as *j* always, or sometimes, or whatever'.[3] It is clear from this that he thinks that the subject should be ampliated to what is at some time *j*. As Street points out, the universal affirmative 'Every *j* is *b*' is taken to mean that all the *j*s (picked out as above) 'are at least once *b*s'.[4] This reading attaches temporal modalities to both the subject '*j*' (which is modified as 'sometimes *j*') and

---

[1] Avicenna, *Salvation* 58.7-8. Translation in Street (2002) p.148.
[2] Ackrill (1973) A6, 1096a15.
[3] Avicenna, *Pointers* 280. Translation in Street (2002) p.134.
[4] Street (2002) p.135.

the predicate '*b*' (which is similarly modified). But Street reminds us that the modalities attaching to subject and predicate may be alethic rather than temporal.[5] If so then an assertoric proposition may mean that every possible *b* is a possible *a*, rather than 'Every sometime *b* is a sometime *a*'. There are deep logical similarities between temporal and alethic modalities, so deep that I shall use a single notation to cover both. In this context I shall read 'a^*' indifferently as 'necessarily *a*' or 'always *a*', and shall read '$a^\dagger$' indifferently as 'possibly *a*' or 'sometimes *a*'. The resulting formalization will therefore be a simplified version of Avicenna's logic.

In this simplified logic, merging alethic and temporal modalities, the meanings of assertoric propositions are stated in Definitions 4.1 to 4.4.

Definition 4.1 (Xa). 'Every *b* is *a*': $\underline{b^\dagger} \rightarrow a^\dagger$, i.e., every possible/sometime *b* is possibly/sometimes *a*.

Definition 4.2 (Xe). 'No *b* is *a*': $b^\dagger \mid a^*$, i.e., no possible/sometime *b* is necessarily/always *a*.

Definition 4.3 (Xi). 'Some *b* is *a*': $b^\dagger \smile a^\dagger$, i.e., some possible/sometime *b* is possibly/at some time *a*.

Definition 4.4 (Xo). 'Some *b* is not *a*': $b^\dagger \nrightarrow a^*$, i.e., not every possible/sometime *b* is necessarily/always *a*.

Given these definitions, universal assertorics do not contradict particular assertorics of opposite quality. Furthermore, assertorics do not in all cases convert as Aristotelian doctrine would have it. Particular affirmative assertorics convert as Aristotle states, but negatives do not convert: if no possible *b* is a necessary *a* it does not follow that no possible *a* is a necessary *b*, and if every possible *b* is a necessary *a* it does not follow that some possible *a* is a necessary *b*. Avicenna argues that 'No man is laughing' is true on his reading – since no possible man is necessarily laughing – but 'No laugher is a man' is not true since some possible laugher is a necessary man.[6] Further, given definitions 4.1 to 4.4, there are no valid assertoric moods in the second figure:

> ...no conclusion follows in the second figure from two absolutes, just as none follows from two possibles....[7]

In other respects Avicenna's account of assertoric syllogistic is orthodox, retaining the standard first- and third-figure moods. Avicenna was aware that the oddities mentioned above could be remedied by reading assertorics *ut nunc*, in which case the standard square of opposition would apply to

---

[5] Street (2002) p.136.

[6] Avicenna, *Pointers* 322. Street (2002) p.143.

[7] Avicenna, *Salvation* 51. Translation in Street (2002) p.146.

assertorics, as would the standard laws of conversion, and the validity of second-figure syllogisms would be restored.[8] Nonetheless he preferred to treat assertorics in the above way, as equivalent to possibility-propositions with ampliated subjects.

Modal logic

Ammonius, and following him Stephanus, had drawn a threefold distinction between propositions of absolute necessity (which may express eternal truths or merely truths so long as the subject exists) and propositions of conditional necessity such as 'You are of necessity walking, as long as you are walking' (which are supposed to be true merely 'as long as that is true which is predicated by the one who says that it is so').

In what appears to be a similar vein, Avicenna distinguishes absolute necessity-propositions (like 'God exists') from conditional necessity-propositions; and within the class of conditional necessities he further distinguishes between those (like 'Man is necessarily a rational body') that are conditional upon the continued existence of the substance that is referred to by the subject-term, and those (like 'All that is moving is changing') that are conditional upon the subject-term's continuing to characterize the substance to which it refers.[9] He calls propositions of the first type 'substantial' [*dhātiyya*] necessity-propositions, and those of the second type 'descriptional' [*wasfiyya*]. Avicenna explains, in reference to the proposition 'Man is necessarily a rational body':

> We do not mean that man is and always will be a rational body, because this is false.... Rather we mean that while he exists as a substance, as a human, he is a rational body.[10]

And he says that the proposition 'All moving things are changing':

> ...is not to be taken to mean that this is so absolutely, nor while the subject exists as a substance, but rather while the substance of the moving thing is moving'.[11]

---

[8] Street (2000) p.132 n.7.
[9] Avicenna (1984) Fourth Method, Chapter 2. Inati p.92.
[10] Avicenna, *Pointers* 265. Translation in Street (2002) p.133.
[11] *Ibid.*

In neither kind of conditional necessity-proposition is the proposition supposed to be true for all time: it is merely supposed to be true for the whole time of the existence of the substance to which the subject-term refers, or for the whole time during which the subject-term characterizes that substance.

Commenting on the distinction between substantial and descriptional propositions, Street says:

> ...the substantial reading corresponds functionally to the divided reading of medieval Western logic, whereas the descriptional reading has no widely used Western analogue.[12]

This is right; but it is also true that the descriptional reading is remarkably similar to Abélard's compound reading, as in 'It is possible for those standing to sit while remaining standing' if this is understood as a *de rebus* compound proposition (as in Table 3.1).

While Avicenna's threefold distinction is in a way isomorphic with that of Ammonius, their accounts differ in two important ways. First, Avicenna notices a range of cases not mentioned by Ammonius: for instance, the restriction 'as long as' could apply to the subject rather than the predicate. Second, while Ammonius places examples like 'Of necessity Socrates is an animal' with the absolute necessities, Avicenna puts them in the group of conditional necessities, dependent in this instance on the continued existence of the subject.

Statements such as 'Every writer moves his fingers, as long as he writes', are called 'general' descriptionals by Rescher.[13] If to such a 'general' proposition we add the words 'but not perpetually', then we have what Rescher calls a 'special' descriptional.[14] We shall examine Avicenna's account of substantial propositions first, then his account of descriptionals.

Substantial propositions

The meanings that Avicenna attaches to substantial necessity- or possibility-propositions are given in Definitions 4.5 to 4.12.

Definition 4.5 (La). 'Every b is necessarily a': $b^{\dagger} \rightarrow a^{*}$.

Definition 4.6 (Le). 'No b is possibly a': $b^{\dagger} \mid a^{\dagger}$.

Definition 4.7 (Li). 'Some b is necessarily a': $b^{\dagger} \smile a^{*}$.

Definition 4.8 (Lo). 'Some b is not possibly a': $b^{\dagger} \nrightarrow a^{\dagger}$.

---

[12] Street (2002) p.133.

[13] Rescher (1967) p.7.

[14] Rescher (1967) p.9.

Definition 4.9 (Ma). 'Every b is possibly a': $b^{\dagger} \rightarrow a^{\dagger}$.
Definition 4.10 (Me). 'No b is necessarily a': $b^{\dagger} \mid a^{*}$.
Definition 4.11 (Mi). 'Some b is possibly a'': $b^{\dagger} \smile a^{\dagger}$.
Definition 4.12 (Mo). 'Some b is not necessarily a': $b^{\dagger} \nrightarrow a^{*}$.

The doctrine that modal propositions ampliate their subjects comes from Alfarabi's (d.950) long commentary on the *Prior Analytics*, which, however, has not survived.[15] The origins of the doctrine presumably lie in Aristotle's distinction (noted in Chapter 1) between saying 'What is b may or not be a' and saying 'What may or may not be b may or may not be a'.

If substantial propositions are understood as in Definitions 4.5 to 4.12, then Le- and Mi-propositions are convertible. The Le-proposition stating that $b^{\dagger} \mid a^{\dagger}$ is convertible by virtue of Axiom 1.2; and consequently Mi-propositions are convertible. Me-propositions are not convertible.[16] La-propositions do not convert (as in Aristotle) to Li-propositions but to Mi-propositions: if every possible b is necessarily a, it does not follow that some possible a is necessarily b, only that some possible a is possibly b. Avicenna's counter-example to the Aristotelian conversion-law here has it that, while all possible writers are necessarily men, it does not follow that some possible man is necessarily writing, only that some possible man is possibly writing.

L/M combinations

Avicenna takes all the standard LML moods to be valid.[17]

Theorem 4.1. The LML moods are valid in Figure 1. Barbara: if $b^{\dagger} \rightarrow a^{*}$ and $c^{\dagger} \rightarrow b^{\dagger}$ then we have the chain '$c^{\dagger} \rightarrow b^{\dagger} \rightarrow a^{*}$', which simplifies by Axiom 1.3 to '$c^{\dagger} \rightarrow a^{*}$'. Celarent: if $b^{\dagger} \mid a^{\dagger}$ and $c^{\dagger} \rightarrow b^{\dagger}$ then we have the chain '$c^{\dagger} \rightarrow b^{\dagger} \mid a^{\dagger}$', which simplifies by Axiom 1.4 to '$c^{\dagger} \mid a^{\dagger}$'. Darii: if $b^{\dagger} \rightarrow a^{*}$ and $c^{\dagger} \leftarrow d \rightarrow b^{\dagger}$ then we have the chain '$c^{\dagger} \leftarrow d \rightarrow b^{\dagger} \rightarrow a^{*}$', which simplifies by Axiom 1.3 to '$c^{\dagger} \leftarrow d \rightarrow a^{*}$'. Ferio: if $b^{\dagger} \mid a^{\dagger}$ and $c^{\dagger} \leftarrow d \rightarrow b^{\dagger}$ then we have the chain '$c^{\dagger} \leftarrow d \rightarrow b^{\dagger} \mid a^{\dagger}$', which simplifies by Axiom 1.4 to '$c^{\dagger} \leftarrow d \mid a^{\dagger}$'.

Theorem 4.1 gives rise to two corollaries, equivalent to it by indirect reduction.

Theorem 4.1a. The LML moods are valid in Figure 2.[18]
Theorem 4.1b. The MMM moods are valid in Figure 3.[19]

---

15 Street (2000) pp.131, 134.
16 Avicenna, *Pointers* 338-340. Street (2002) p.145.
17 Street (2002) p.160.
18 Avicenna, *Salvation* 66, 5-8. Street (2002) p.153.
19 Avicenna, *Salvation* 60, 64-67. Street (2002) p.150.

Theorem 4.2. The LML moods are valid in Figure 3. Datisi: if $c^{\dagger} \rightarrow a^{*}$ and $c^{\dagger} \leftarrow d \rightarrow b^{\dagger}$, then we have the chain '$b^{\dagger} \leftarrow d \rightarrow c^{\dagger} \rightarrow a^{*}$', which simplifies by Axiom 1.3 to '$b^{\dagger} \leftarrow d \rightarrow a^{*}$'. Disamis: if $c^{\dagger} \leftarrow d \rightarrow a^{*}$ and $c^{\dagger} \rightarrow b^{\dagger}$ then by similar reasoning $b^{\dagger} \leftarrow d \rightarrow a^{*}$. Ferison: if $c^{\dagger} \mid a^{\dagger}$ and $c^{\dagger} \leftarrow d \rightarrow b^{\dagger}$ then we have the chain '$b^{\dagger} \leftarrow d \rightarrow c^{\dagger} \mid a^{\dagger}$', which simplifies by Axiom 1.4 to '$b^{\dagger} \leftarrow d \mid a^{\dagger}$'. Bocardo: if $c^{\dagger} \leftarrow d \mid a^{\dagger}$ and $c^{\dagger} \rightarrow b^{\dagger}$ then by similar reasoning $b^{\dagger} \leftarrow d \mid a^{\dagger}$.

This too has two corollaries, equivalent to it by indirect reduction.

Theorem 4.2a. The MMM moods are valid in Figure 1.[20]

Theorem 4.2b. The MLL moods are valid in Figure 2.[21]

He also accepts the LLL moods in all figures.[22]

Theorem 4.3. The LLL moods are valid in Figure 1. Barbara: if $b^{\dagger} \rightarrow a^{*}$ and $c^{\dagger} \rightarrow b^{*}$ then, interpolating Axioms 1.7 and 1.8, we have the chain '$c^{\dagger} \rightarrow b^{*} \rightarrow b \rightarrow b^{\dagger} \rightarrow a^{*}$', which simplifies by Axiom 1.3 to '$c^{\dagger} \rightarrow a^{*}$'. Celarent: if $b^{\dagger} \mid a^{\dagger}$ and $c^{\dagger} \rightarrow b^{*}$ then, by a similar process but using Axiom 1.4, $c^{\dagger} \mid a^{\dagger}$. Darii: if $b^{\dagger} \rightarrow a^{*}$ and $c^{\dagger} \leftarrow d \rightarrow b^{*}$, then similarly $c^{\dagger} \leftarrow d \rightarrow a^{*}$. Ferio: if $b^{\dagger} \mid a^{\dagger}$ and $c^{\dagger} \leftarrow d \rightarrow b^{*}$ then $c^{\dagger} \leftarrow d \mid a^{\dagger}$.

Theorem 4.4. The LLL moods are valid in Figure 2. Cesare: if $b^{\dagger} \mid a^{\dagger}$ and $c^{\dagger} \rightarrow a^{*}$ then, interpolating Axioms 1.7 and 1.8, we have the chain '$b^{\dagger} \mid a^{\dagger} \leftarrow a \leftarrow a^{*} \leftarrow c^{\dagger}$', which simplifies by Axiom 1.4 to '$b^{\dagger} \mid c^{\dagger}$'. Camestres: if $b^{\dagger} \rightarrow a^{*}$ and $c^{\dagger} \mid a^{\dagger}$ then by similar reasoning we have the chain '$c^{\dagger} \mid a^{\dagger} \leftarrow a \leftarrow a^{*} \leftarrow b^{\dagger}$', which simplifies by Axiom 1.4 to '$c^{\dagger} \mid b^{\dagger}$'. Festino: if $b^{\dagger} \mid a^{\dagger}$ and $c^{\dagger} \leftarrow d \rightarrow a^{*}$ then $c^{\dagger} \leftarrow d \rightarrow a^{*} \rightarrow a \rightarrow a^{\dagger} \mid b^{\dagger}$, and so $c^{\dagger} \leftarrow d \mid b^{\dagger}$. Baroco: if $b^{\dagger} \rightarrow a^{*}$ and $c^{\dagger} \leftarrow d \mid a^{\dagger}$ then $c^{\dagger} \leftarrow d \mid a^{\dagger} \leftarrow a \leftarrow a^{*} \leftarrow b^{\dagger}$, and so $c^{\dagger} \leftarrow d \mid b^{\dagger}$.

Theorem 4.5. The LLL moods are valid in Figure 3. Datisi: if $c^{\dagger} \rightarrow a^{*}$ and $c^{\dagger} \leftarrow d \rightarrow b^{*}$ then, interpolating Axioms 1.7 and 1.8, we have the chain '$b^{\dagger} \leftarrow b \leftarrow b^{*} \leftarrow d \rightarrow c^{\dagger} \rightarrow a^{*}$' which simplifies to '$b^{\dagger} \leftarrow d \rightarrow a^{*}$'. Disamis: if $c^{\dagger} \leftarrow d \rightarrow a^{*}$ and $c^{\dagger} \rightarrow b^{*}$ then by similar reasoning $b^{\dagger} \leftarrow b \leftarrow b^{*} \leftarrow c^{\dagger} \leftarrow d \rightarrow a^{*}$ and so $b^{\dagger} \leftarrow d \rightarrow a^{*}$. Ferison: if $c^{\dagger} \mid a^{\dagger}$ and $c^{\dagger} \leftarrow d \rightarrow b^{*}$ then we have the chain '$b^{\dagger} \leftarrow b \leftarrow b^{*} \leftarrow d \rightarrow c^{\dagger} \mid a^{\dagger}$', and so $b^{\dagger} \leftarrow d \mid a^{\dagger}$. Bocardo: if $c^{\dagger} \leftarrow d \mid a^{\dagger}$ and $c^{\dagger} \rightarrow b^{*}$ then $b^{\dagger} \leftarrow b \leftarrow b^{*} \leftarrow c^{\dagger} \leftarrow d \mid a^{\dagger}$ and so $b^{\dagger} \leftarrow d \mid a^{\dagger}$.

Notice that in order to prove the validity of the LLL moods (as opposed to the LMM and MLM moods) in each of the figures we have to use interpolation.

Comparing Table 4.1 with Table 1.6, it is clear that Avicenna's system properly includes Aristotle's L/M system. This comparison is displayed in Figure 4.1.

---

[20] Avicenna, *Salvation* 60, 64-67. Street (2002) p.150.

[21] Street (2002) p.153.

[22] Street (2002) pp.147-148.

Table 4.1 Avicenna's system: L/M combinations

Figure	*Combinations*[23]
1	LLL, LML, LMM, MLM, MMM
2	LLL, LML, LMM, MLL, MLM
3	LLL, LML, LMM, MLM, MMM

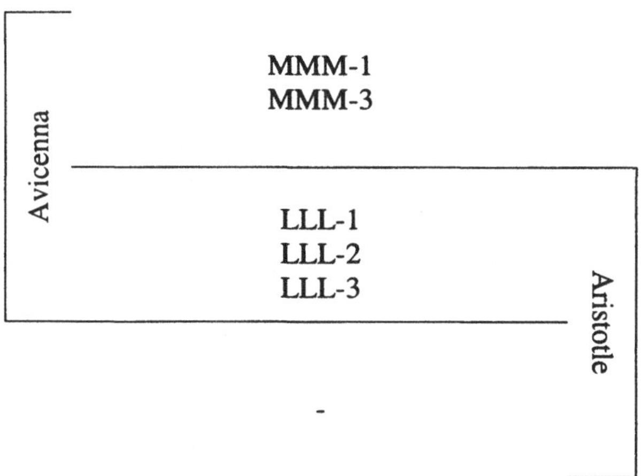

Figure 4.1 Avicenna's L/M system and Aristotle's system

Each of the above theorems has two corollaries by indirect reduction.
[**Theorem 4.3a.** The LMM moods are valid in Figure 2.]
Theorem 4.3b. The MLM moods are valid in Figure 3.[24]
[**Theorem 4.4a.** The LMM moods are valid in Figure 1.]
[**Theorem 4.4b.** The LMM moods are valid in Figure 3.]
[**Theorem 4.5a.** The MLM moods are valid in Figure 1.]
[**Theorem 4.5b.** The MLM moods are valid in Figure 2.]

[23] Missing from Appendix 4 in Street (2002) p.160 are LMM-1, LMM-2, MLM-2, LMM-3.
[24] Street (2002) p.153.

L/X/M combinations

The validity of syllogisms with mixed assertoric-necessity-premises can be inferred from the cases in which there are mixed possibility-necessity-premises. Wherever LM or ML or MM premises yield a conclusion, the corresponding LX or XL or XX or MX or XM premises yield the same conclusion; and wherever premises yield an M-conclusion they also yield the corresponding X-conclusion. Thus we have the following correspondences from Table 4.1 (omitting the LLL moods and their equivalents namely LMM-1, MLM-1, LMM-2, MLM-2, LMM-3 and MLM-3).

Table 4.2 Correspondences between L/M and L/X/M moods

LML-1	LXL-1
LMM-1	LXX-1, LXM-1, LMX-1
MLM-1	XLM-1, MLX-1, XLX-1
MMM-1	XXX-1, XXM-1, XMX-1, XMM-1, MXX-1, MXM-1
LML-2	LXL-2
LMM-2	LXX-2, LXM-2, LMX-2
MLL-2	XLL-2
MLM-2	XLX-2, XLM-2, MLX-2
LML-3	LXL-3
LMM-3	LXX-3, LXM-3, LMX-3
MLM-3	XLX-3, XLM-3, MLX-3
MMM-3	XXX-3, XXM-3, XMX-3, XMM-3, MXX-3, MXM-3

These fall into two groups whose members are mutually equivalent.

1. LXL-1, LXL-2, XXX-3, XXM-3, XMX-3, XMM-3, MXX-3, MXM-3
2. LXL-3, XLL-2, XXX-1, XXM-1, XMX-1, XMM-1, MXX-1, MXM-1.

Avicenna is aware of some of these cases. He takes the LXL moods of Figure 1 to be valid.[25] His semantics entitles him to dismiss Alexander's proposed counter-examples to Barbara LXL,[26] on the grounds that in one case the minor premise ('Everything that is possibly moving is possibly human') is false, and in the others the major ('Everything that is possibly literate necessarily has knowledge' and 'Everything that is

[25] Street (2002) p.148.
[26] See above p.28.

possibly walking is necessarily moving') is false. In the second figure he accepts both versions (XLL and LXL) of each of the standard moods.[27] His response to Alexander's putative counter-example to Baroco[28] should be to reject the major premise ('Everything possibly moving is possibly an animal'). In Figure 3 he accepts all LXL moods. At the same time, the Aristotelian Disamis XLL is invalid: if $c^{\dagger} \smile a^{\dagger}$ and $c^{\dagger} \rightarrow b^{*}$, then it does not follow that $b^{\dagger} \smile a^{*}$; otherwise we could argue that if $c^{\dagger} \rightarrow b^{*}$ and $b^{\dagger} \mid a^{*}$ then $c^{\dagger} \mid a^{\dagger}$; where all that can be concluded from those premises is $c^{\dagger} \mid a^{*}$. As Avicenna says, 'the truth is that the conclusion follows the major'.[29] Avicenna recognizes no valid mixed X/M mixtures in the second figure, but thinks that all standard first- or third-figure moods are valid in either configuration (XM or MX) to an M conclusion.[30]

Table 4.3 Avicenna's system: L/X/M combinations

Figure	*Combinations*[31]
1	LXL, XXM, XMX, XMM, MXM, MXX
2	LXL, XLL
3	LXL, XXM, XMX, XMM, MXX, MXM

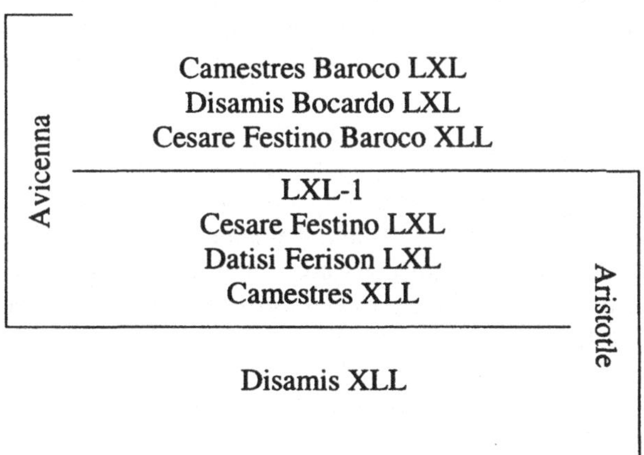

Figure 4.2 Avicenna's L/X/M system and Aristotle's system

[27] Street (2002) p.148.
[28] See above p.29.
[29] Avicenna, *Salvation* 59.12-13. Translation in Street (2002) p.149.
[30] Avicenna, *Salvation* 61-67. Street (2002) pp.150-151.
[31] Missing from Appendix 4 in Street (2002) p.160 are XXM-1, XMX-1, MXX-1, XXM-3, XMX-3, MXX-3.

Avicenna's L/X/M system includes Abélard's, along with the LXL and XLL moods in the second figure (and their equivalents). Figure 4.2 compares the groups of L/X/M syllogisms accepted by Aristotle and Avicenna respectively. It should be noted that, where Avicenna and Aristotle agree on a syllogism's validity, they can also agree on the method of demonstrating its validity. Aristotle reduces Cesare and Festino LXL, and Camestres XLL, to the first figure by converting universal negative necessity-propositions. He reduces Datisi and Ferison LXL to the first figure by converting particular affirmative assertoric propositions. Both these processes are acceptable according to Avicenna's semantics.

Descriptional Propositions

Avicenna was not the first to discuss those necessity-propositions that include a restriction of the form 'as long as…'. Aristotle had set the topic going with his 'Everything that is, is by necessity, as long as it is'. Ammonius had drawn attention to the difference between the case in which what restricts the proposition is the subject's continued existence, and that in which the restriction is based on the continued application of the predicate. Abélard had made further subtle distinctions. What none of these thinkers had done was to construct a theory articulating the inferential capacities of these propositions, their logic. It was Avicenna who achieved the first results in this field.

In representing descriptional propositions I shall use two new notations: 'a/b' (read 'a-if-b' or 'a-when-b') and 'ab' (read 'a and b' or 'a while b'). In order to distinguish descriptionals from the ordinary (assertoric) X-form, I designate specials by 'X_1' and generals by 'X_2'. Then the truth-requirements for universal specials are stated in Definitions 4.13 and 4.14. (For simplicity's sake I will use modal rather than temporal language.)

Definition 4.13 (X_1a). 'Every b is necessarily a-if-b (but none is necessarily a)': $b^{\dagger} \rightarrow (a/b)^{*}$ and $b^{\dagger} \mid a^{*}$.

Definition 4.14 (X_1e). 'No b is possibly a-and-b (but some is possibly a)': $b^{\dagger} \mid (ab)^{\dagger}$ and $b^{\dagger} \smile a^{\dagger}$.

As examples of such propositions we may take 'All of those who are possibly/sometimes writing are necessarily/always moving as long as they are writing, but none of them are necessarily/always writing', and 'None of those who are possibly/sometimes sitting are possibly/ever standing so long as they are sitting, but some of them are possibly/sometimes standing'. For generals, as opposed to specials, we drop the second conjunct in these definitions.

Definition 4.15 (X_2a). 'Every b is necessarily a-if-b': $b^{\dagger} \rightarrow (a/b)^{*}$

Definition 4.16 (X_2e). 'No *b* is possibly *a*-and-*b*': $b^\dagger \mid (ab)^\dagger$.

Definition 4.17 (X_2i).'Some *b* is necessarily *a*-if-*b*': $b^\dagger \smile (a/b)^*$.

Definition 4.18 (X_2o). 'Not every *b* is possibly *a*-and-*b*': $b^\dagger \nrightarrow (ab)^\dagger$.

Clearly the contradictory of an X_2 is not an X_2. We need another label for the contradictories of X_2 propositions; let it be X_3. Then we can adopt the following definitions:

Definition 4.19 (X_3a).'Every *b* is possibly *a*-and-*b*': $b^\dagger \rightarrow (ab)^\dagger$.

Definition 4.20 (X_3e).'No *b* is necessarily *a*-if-*b*': $b^\dagger \mid (a/b)^*$.

Definition 4.21 (X_3i).'Some *b* is possibly *a*-and-*b*': $b^\dagger \smile (ab)^\dagger$.

Definition 4.22 (X_3o).'Not every *b* is necessarily *a*-if-*b*': $b^\dagger \nrightarrow (a/b)^*$.

We now state some axioms that will govern these new notions.

Axiom 4.1. $(ab)^\dagger \rightarrow a^\dagger$. What is possibly *a*-and-*b* is possibly *a*.

Axiom 4.2. $a^* \rightarrow (a/b)^*$. What is necessarily *a* is necessarily *a*-if-*b*.

Axiom 4.3. If $c \rightarrow b^*$ and $c \rightarrow (a/b)^*$ then $c \rightarrow a^*$. What is necessarily *b*, and necessarily *a*-if-*b*, is necessarily *a*.

Axiom 4.4. If $c \rightarrow b^*$ and $c \mid (ab)^\dagger$ then $c \mid a^\dagger$. What is necessarily *b*, and not possibly *a*-and-*b*, is not possibly *a*.

Axiom 4.5. If $c \rightarrow b^\dagger$ and $c \rightarrow (a/b)^*$ then $c \rightarrow (ab)^\dagger$. What is possibly *b*, and necessarily *a*-if-*b*, is possibly *a*-and-*b*.

Axiom 4.6. If $d \rightarrow (a/b)^*$ and $d \rightarrow (b/c)^*$ then $d \rightarrow (a/c)^*$. What is necessarily *a*-if-*b*, and necessarily *b*-if-*c*, is necessarily *a*-if-*c*.

Axiom 4.7. If $d \mid (ab)^\dagger$ and $d \rightarrow (b/c)^*$ then $d \mid (ac)^\dagger$. What is not possibly *a*-and-*b* and necessarily *b*-if-*c*, is not possibly *a*-and-*c*.

Axiom 4.8. If $d \rightarrow (ab)^\dagger$ and $d \rightarrow (c/b)^*$ then $d \rightarrow (ac)^\dagger$. What is possibly *a*-and-*b* and necessarily *c*-if-*b*, is possibly *a*-and-*c*.

X_2- and X_3-propositions obey laws of conversion as follows. 'No *b* is possibly *a*-and-*b*' converts to 'No *a* is possibly *b*-and-*a*'. Proof: given Axiom 4.1, if $b^\dagger \mid (ab)^\dagger$ then nothing is $(ab)^\dagger$, and so nothing is $(ba)^\dagger$, in which case $a^\dagger \mid (ba)^\dagger$. 'Some *b* is possibly *a* and *b*' converts to 'Some *a* is possibly *b* and *a*'. Proof: if $\sim[(ba)^\dagger \mid a^\dagger]$ then $\sim[b^\dagger \mid (ab)^\dagger]$.

X_2i and X_3e propositions are not convertible. Consider the X_2i: it may be that some possible Persian-speaker is necessarily speaking-if-speaking-Persian, yet no possible speaker is necessarily speaking-Persian-if-speaking. (Even one who is speaking Persian is not necessarily speaking-Persian-if-speaking.)

I give sample proofs for some of the implication-relations that hold between L-, X_2-, X_3- and M-propositions. (i) La implies X_2a; (ii) X_2a implies X_3a; (iii) X_3a implies Ma. Proof of (i): if $b^\dagger \rightarrow a^*$ then $b^\dagger \rightarrow (a/b)^*$ (from Axiom 4.2). (If, as this implication shows, X_2 propositions are weaker than L propositions, then in order to construct a counter-example to Theophrastus' *peiorem* rule it would be sufficient to show that an X_2LL syllogism is valid. As we are about to see, this is precisely what Avicenna does.) Proof of (ii): if $b^{\dagger} \rightarrow (a/b)^{*}$ then $b^{\dagger} \rightarrow (ab)^{\dagger}$, by Axiom 4.5. Proof of (iii): if $b^{\dagger} \rightarrow (ab)^{\dagger}$ then $b^{\dagger} \rightarrow a^{\dagger}$, by Axiom 4.1. These logical relationships are exhibited in Figure 4.3. Relations of implication are indicated by arrows; relations of contradiction are indicated by light lines.

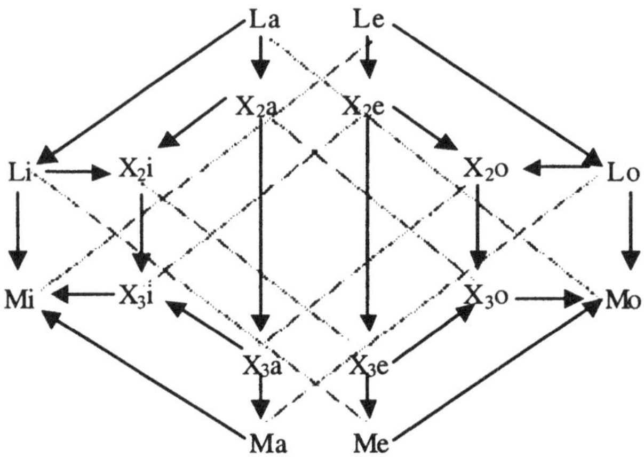

Figure 4.3 Relations among descriptionals

Avicenna's discussion of syllogisms with descriptional premises begins with a reference to Theophrastus's *peiorem* rule, in so far as it relates to the modality of the premises and conclusion.[32] He does not accept the rule in relation to modality. As we have seen, he agrees with Aristotle in accepting Barbara LXL while rejecting Barbara XLL. However, Avicenna believes that under a suitable interpretation of the assertoric major the first-figure XLL moods are valid. He states that in the first figure if the minor is a necessity-proposition and the major an assertoric asserting the predicate of the subject 'as long as the subject is qualified by that with which it is qualified',[33] then firstly the syllogism cannot have true premises; but secondly if the major is understood in a more general sense, the premises can be true together, and a necessity-conclusion follows.

His reason for saying that the premises cannot be true together is that:

[32] Avicenna (1984) Seventh Method, Chapter 5; p.137.

[33] Inati translation.

> ...if we say 'Every *c* is by necessity *b*', and then we say 'Every *b* is qualified as *a*, as long as it is qualified as *b*, and not always', we judge that all that which is qualified as *b* is qualified thus, only at a certain time, and not always.[34]

Thus he requires that the major be a special descriptional stating both that $b^\dagger \rightarrow (a/b)^*$ and that $b^\dagger \mid a^*$. He takes this to be incompatible with the minor stating that $\underline{c}^\dagger \rightarrow b^*$, because the qualifying phrase in the major implies that no *b* is necessarily/always *b*, whereas the minor implies that some *b* is necessarily/always *b*. Formally, if $b^\dagger \rightarrow (a/b)^*$ and $b^\dagger \mid a^*$ then not $(\underline{c}^\dagger \rightarrow b^*)$. Proof: if $\underline{c}^\dagger \rightarrow b^*$ then some possible *b* is necessarily *b* because we may form the chain '$b^\dagger \leftarrow b^* \leftarrow \underline{c}^\dagger \rightarrow b^*$', which simplifies to '$b^\dagger \leftarrow \underline{c}^\dagger \rightarrow b^*$'. On the other hand, if $b^\dagger \rightarrow (a/b)^*$ and $b^\dagger \mid a^*$ then no possible *b* is necessarily *b* because whatever is necessarily *a*-if-*b* is necessarily *a* if it is necessarily *b* (Axiom 4.3); so if $b^\dagger \rightarrow (a/b)^*$ and $b^\dagger \mid a^*$ then $b^\dagger \mid b^*$. Avicenna has noticed that Barbara X_1LL has mutually incompatible premises, the major implying, what the minor denies, that the middle term is *per accidens*. We shall later find Richard Campsall noting another syllogism with incompatible premises, where the major implies, and the minor denies, that the middle term is *per se* (Theorem 7.26).

By contrast with an X_1-major, an X_2-major stating merely that $b^\dagger \rightarrow (a/b)^*$ is perfectly compatible with the minor. Avicenna believes that Barbara X_2LL is valid. This is correct, and can be stated as a more general theorem.

Theorem 4.6. The X_2LL moods are valid in Figure 1. Barbara: if $\underline{b}^\dagger \rightarrow (a/b)^*$ and $\underline{c}^\dagger \rightarrow b^*$ then by interpolation we may form the chain '$\underline{c}^\dagger \rightarrow b^* \rightarrow b \rightarrow b^\dagger \rightarrow (a/b)^*$', which simplifies to '$\underline{c}^\dagger \rightarrow (a/b)^*$', and this implies that if $\underline{c}^\dagger \rightarrow b^*$ then $\underline{c}^\dagger \rightarrow a^*$ (by Axiom 4.3). Celarent: if $b^\dagger \mid (ab)^\dagger$ and $c^\dagger \rightarrow b^*$ then by similar reasoning, but using Axiom 4.4, $c^\dagger \mid a^\dagger$. Darii: if $b^\dagger \rightarrow (a/b)^*$ and for some '*d*' $c^\dagger \leftarrow \underline{d} \rightarrow b^*$, then for some '*d*', $c^\dagger \leftarrow \underline{d} \rightarrow a^*$ by Axiom 4.3. Ferio: analogously. Avicenna's position should be contrasted with that of Sosigenes (noted in Chapter 2). Sosigenes thought that Barbara LXL needed an unconditional necessity-major and a conditional necessity-conclusion, whereas Avicenna thought that Barbara XLL would be valid if it had a conditional major and an unconditional conclusion.

The two corollaries of Theorem 4.6 by indirect reduction are as follows.

Theorem 4.6a. The X_2MM moods are valid in Figure 2. Avicenna does not identify any syllogisms containing general descriptionals in Figures 2 and 3. However his commentator Tūsī does:

---

[34] Avicenna (1984) pp.99-100.

> Possibly all men move and No sleeping man moves while sleeping… produces Possibly no man is sleeping.[35]

This is a Cesare X_2MM: if $b^\dagger \mid (ab)^\dagger$ and $c^\dagger \to a^\dagger$ then $c^\dagger \mid b^*$. *a*: movers; *b*: sleepers; *c*: men.

[**Theorem 4.6b.** The MLX_3 moods are valid in Figure 3.]

Also provable is:

[**Theorem 4.7.** The $X_2X_2X_2$ moods are valid in Figure 1.] Barbara: if $b^\dagger \to (a/b)^*$ and $c^\dagger \to (b/c)^*$, the minor premise can be replaced first by ‘$c^\dagger \to (bc)^\dagger$’ (by virtue of Axiom 4.5) and then by ‘$c^\dagger \to b^\dagger$’ (by virtue of Axiom 4.1), so that we may form the chain ‘$c^\dagger \to b^\dagger \to (a/b)^*$’, which simplifies to ‘$c^\dagger \to (a/b)^*$’; and this together with the minor premise entails that $c^\dagger \to (c/a)^*$ (by Axiom 4.6). Celarent: if $b^\dagger \mid (ab)^\dagger$ and $c^\dagger \to (b/c)^*$, the minor premise can be replaced first by ‘$c^\dagger \to (bc)^\dagger$’ (by Axiom 4.5) and then by ‘$c^\dagger \to b^\dagger$’ (by Axiom 4.1), so that we may form the chain ‘$c^\dagger \to b^\dagger \mid (ab)^\dagger$’, which simplifies to ‘$c^\dagger \mid (ab)^\dagger$’; and this together with the minor premise entails that $c^\dagger \mid (ac)^\dagger$ (by Axiom 4.7). Darii and Ferio can be proved in similar ways.

The two corollaries are as follows.

Theorem 4.7a. The $X_2X_3X_3$ moods are valid in Figure 2. Avicenna's commentator Ṭūsī gives an instance:

> An example is: The sitting may not move his hand at some times while sitting, and The writer moves his hand at all times while writing, which produces The sitter may not be writing at all times of his sitting.[36]

In our interpretation this is a Baroco $X_2X_3X_3$: if $b^\dagger \to (a/b)^*$ and $c^\dagger \nrightarrow (a/c)^*$ then $c^\dagger \nrightarrow (b/c)^*$. *a*: hand-movers; *b*: writers; *c*: sitters.

[**Theorem 4.7b.** The $X_3X_2X_3$ moods are valid in Figure 3.]

We also have the following theorem.

[**Theorem 4.8.** The $X_2X_3X_3$ moods are valid in Figure 1.] Barbara $X_2X_3X_3$: if $b^\dagger \to (a/b)^*$ and $c^\dagger \to (bc)^\dagger$, then by interpolation of Axiom 4.1 we may form the chain ‘$c^\dagger \to (a/b)^\dagger \to b^\dagger \to (a/b)^*$, which simplifies to ‘$c^\dagger \to (a/b)^*$; and this, together with the minor premise entails ‘$c^\dagger \to (ac)^\dagger$ (by Axiom 4.8). Celarent, Darii and Ferio $X_2X_3X_3$ can be proved similarly.

Its two corollaries are as follows.

[**Theorem 4.8a.** The $X_2X_2X_2$ moods are valid in Figure 2.]

Theorem 4.8b. $X_2X_3X_3$ Figure 3. Ṭūsī gives an example:

---

[35] *Ibid.*

[36] Ṭūsī *ad* Seventh Method, Chapter 6. Translation in Street (n.d.).

> ...so we say: All writers are awake and using a pen while they write; and we do not say: Some who are awake use a pen while they are awake, rather [Some who are awake use a pen] at some times that they are awake.[37]

This is a Darapti $X_2X_2X_3$: if $\underline{c}^{\dagger} \rightarrow (a/c)^{*}$ and $\underline{c}^{\dagger} \rightarrow (b/c)^{*}$ then $b^{\dagger} \smile (ab)^{\dagger}$. *a*: pen-users; *b*: those who are awake; *c*: writers.

Table 4.4 Descriptional syllogisms

Figure	*Combinations*
1	X_2LL, $X_2X_2X_2$, $X_2X_3X_3$
2	X_2MM, $X_2X_3X_3$, $X_2X_2X_2$
3	MLX_3, $X_3X_2X_3$, $X_2X_3X_3$

Given the implication relations shown in Figure 4.3, the validity of some syllogisms follows from that of others having either weaker premises or stronger conclusions. Thus given any syllogism with an L conclusion we have a corresponding syllogism with X_2, X_3 and M conclusions; and given any syllogism with an M premise we have corresponding syllogisms with X_3, X_2 and L premises.

Summary

Avicenna's modal logic has three distinctive features. One is his ampliation of the subjects of modal propositions. From a theoretical point of view, this device can be seen as an alternative to the *de rebus* semantics advocated by Abélard, an alternative that redeems at least some of the modal conversion-principles rejected by Abélard (Le- and Mi-conversion) along with all but one (Disamis XLL) of the Aristotelian mixed necessity-syllogisms. At the same time it delivers a new batch of unAristotelian syllogisms (Camestres and Baroco LXL, Cesare Festino and Baroco XLL). As an interpretive strategy, it solves one problem while creating others.

The second distinctive feature of Avicenna's modal logic is his identification of assertorics with possibility-propositions. This enables him to treat the L/M layer as primary, with L/M/X syllogisms secondary. From a systematic point of view, this is a gain; but it also entails distortions in the logic of assertoric propositions: we lose some aspects of the traditional square of opposition, along with the convertibility of universal negatives and the validity of the second figure syllogisms. His logic of substantive

[37] Ṭūsī *ad* Seventh Method, Chapter 7. Translation in Street (n.d.).

modals can be based on the pure necessity-syllogisms in Figures 1, 2 and 3, and the pure possibility-syllogisms in Figures 1 and 3. (In view of the convertibility of Le- and Mi-propositions, Cesare, Camestres and Festino LLL, along with Datisi, Disamis and Ferison MMM, can be omitted from this base.)

The third distinctive element concerns the logic of descriptionals. Here Avicenna has identified a form of proposition – or rather, two forms of proposition – neither of which occurs in Aristotle's modal syllogistic. These two forms are the special (X_1) and general (X_2) descriptional. Either a special or a general together with an L-minor in Barbara validly yields an L-conclusion. One of the forms is incompatible with the L-minor; the other is not. Neither of these forms is a simple X; so it would not be right to say that Avicenna has discovered a way in which Barbara XLL is valid. He has, however, discovered a counter-example to Theophrastus's rule, seeing that L-propositions are stronger than X_2-propositions.

Chapter 5

Averröes

Averröes (Ibn Rushd d.1198) is an altogether different type from Avicenna, a true commentator, who labours to find ways of saving the truth of Aristotle's words. Notwithstanding the significance of his commentaries, it is not them but one of his late essays that will be our focus in this chapter. Averröes worked on the interpretation of the *Prior Analytics* at various stages of life, and towards the end of his life he wrote an essay on modal syllogistic, familiar to the Latin world as the *Quaesitum* on mixed syllogisms.[1] This late work is aptly summarized by Elamrani-Jamal:

> Returning at the end of his career to a literal exegesis of Aristotle's proposals... Ibn Rušd [sic] proposes a theory of the modalization of terms... in order to save the literalness of Aristotle's statements... about the mode of the conclusion in mixed syllogisms.[2]

Per se necessities

In a lengthy but important passage from this late work, Averröes classifies propositions according to the types of terms occurring in them:

> It is useful for you to know that assertoric propositions have assertoric terms, necessary propositions necessary terms. [1] By 'necessary terms' I mean that the term is one *per se*, and these [propositions] are composed of a subject and an essential predicate of that subject, or a subject and a connected accident inhering in the subject. [2] Those [propositions] with assertoric terms are composed of denominative terms, because sometimes they are present in the denominated thing and sometimes they are absent, but so long as one of these denominations is found in a subject the consequent denomination must necessarily be found in the predicate, for example, when we say that whatever is walking is moving, when walking is found in actuality so is motion and when it is absent so is

[1] For an account of Averröes's progress towards this 'final state', see Elamrani-Jamal (1995). Lagerlund (2000) pp.32-35 has emphasized the importance of Averröes's text for Western logicians of the thirteenth century.

[2] Elamrani-Jamal (1995) p.74.

> motion.[3] These are simple assertoric propositions, which lack any specific time, and these are what he is primarily talking about in this book, and their subject and predicate alike are one *per accidens*. [3] And there exists another kind of proposition that is partly assertoric and partly necessary. This is where the subject is composed of a substance and a different denomination, and the predicate is composed of the subject of the denomination and from a fundamental essential denomination. The subject here is one *per accidens* and the predicate is one *per se*, for example, when we say that everything walking is an animal. And this is assertoric on account of the denomination of the denominated subject, and necessary on account of the predicate of the denomination. And walking, when it occurs, signifies an animal by discontinuous signification, but for the times at which walking is present in it. The subject of walking implies being an animal always, because the subject of walking and what is denoted by that is necessarily an animal. And this proposition is partly necessary and partly assertoric – assertoric *per se* and necessary *per accidens*. [4] A proposition, that conversely has a necessary subject and a predicate of assertoric matter, is just [*simpliciter*] assertoric, and is not necessary *per accidens*. This is a temporal assertoric, I say, where the subject implies the predicate for a specified time, and necessity is not found in it, only a connection of the predicate and subject merely for that time.[4] The characteristic of this [proposition] is that the predicate is not connected to the subject for all times at which that subject exists, but only for a certain specified time. And so (as Aristotle says) syllogisms in the sciences are not constructed from this type of assertoric, nor are they composed solely of contingent propositions. A minor, or a major, that is just [*simpliciter*] an assertoric that is not for a specified time, is not to be joined with a necessary proposition. That being so, it is not to be denied that [an assertoric minor] may be joined with a necessary major where it is assertoric *per se* and necessary *per accidens*. And this necessarily requires that there is a conclusion that concludes from those in matter that is necessary *per accidens* by virtue of the fact that in them is found a *per accidens* necessary *dici de omni*.... But when the minor depending on a necessary major is likewise necessary, the conclusion is necessary *per se*.[5]

The notion that some terms are 'one *per se*', others 'one *per accidens*', comes from Aristotle's *Posterior Analytics*, where 'walking' and 'white' are given as instances of *per accidens* terms, as contrasted with the *per se* term 'animal'.[6] We encountered this distinction in Chapter 1, where we defined a *per se* term '*a*' as a term such that $\Box a \rightarrow a^*$ (Definition 1.6). Our earlier distinction opposed *per se* terms to terms that were either mixed

---

[3] Recall Alexander's denial, noted in Chapter Two (p.27 above) that this proposition is unconditionally necessary.

[4] Averröes (1562) Q.4 ch.3, 83M-84B.

[5] Averröes (1562) Q.4 ch.3, 84F.

[6] Aristotle (1975) A4, 73b4-9.

(like 'white') or *per accidens* (like 'walking'). It seems that Averröes groups mixed terms like 'white' with *per accidens* terms. What is novel in Averröes's thinking here is the notion that different classes of modal syllogism are differentiated by the type of terms occurring in them – necessity-syllogisms containing necessary terms, assertoric syllogisms containing assertoric terms. Averröes claims to find this notion in Aristotle,[7] appealing to the following text:

> Since to belong and to belong of necessity and to be possible to belong are different (for many things belong, but nevertheless not of necessity, while others neither belong of necessity nor belong at all, but it is possible for them to belong), it is clear that there will also be different deductions of each and that their terms will not be alike: rather, one deduction will be from necessary terms, one from terms which belong, and one from possible terms.[8]

It must be said that Averroes's reading of this text is not the only possible one, and that it appears to conflict with Aristotle's ideas elsewhere. One reason for thinking that Averröes has misread this passage is that Aristotle's non-modal syllogistic appears to be a formal system that is applicable to all terms, including *per se* terms.[9] Nonetheless, Averröes's reading remains historically important.

At first sight, it appears that Averröes is here constructing a division of all inclusions, based on the character (*per se* or *per accidens*) of their subject and predicate. Either [1] both terms are *per se*, or [2] neither is *per se*, or [3] the predicate is *per se* and the subject is not, or [4] the subject is *per se* and the predicate is not. Propositions of type [1] are said to be *per se* necessary. Type [3] are described as *per se* assertoric and *per accidens* necessary. Type [4] are *per se* assertoric but not *per accidens* necessary (nor, obviously *per se* necessary). Propositions of type [2] are said to be simple assertorics, though it would be more in keeping with the overall tenor of the division to say that they are *per accidens* assertorics.

However, this reading will not stand. Averröes's examples ('A subject and an essential predicate of that subject', 'Everything walking is moving', 'Everything walking is an animal') make it obvious that he has *necessary* inclusions in mind in speaking of propositions of types 1, 2 and 3. Perhaps he intends a division of all inclusions, first into necessary then into non-necessary (type 4) inclusions, where necessary inclusions either have both

---

[7] Aristotle (1989) A8, 29b28ff.

[8] Smith translation.

[9] *Per se* terms are used in counter-examples to assertoric syllogisms at Aristotle (1989) A4, 26a2-9, 36-39, 26b3-14, 21-25, and in many other places.

terms *per se* (type 1), or neither term *per se* (type 2), or only the predicate *per se* (type 3). There are two problems with this reading. Firstly, it provides no place for necessary inclusions in which only the subject is *per se*. Secondly, it gathers all non-necessary inclusions – even those in which one or other terms is *per se* – together under type 4.

I propose a third reading that takes into account Averröes's larger purpose in this passage, which is to solve the logical problem of the two Barbaras, by means of an analysis that incorporates metaphysical as well as purely logical elements. He believes that his classification of inclusion-propositions based on the *per se* or *per accidens* status of their terms will provide him with a way of determining the validity or invalidity of various types of modal syllogism. Such a classification will be significant because of its recognition of a class of *per se* necessary propositions. Such propositions occupy a fundamental place in Aristotelian essentialism, since they express the central class of metaphysical truths, namely those that arise from the subordination of a species to its genus, its differentia, its necessary peculiarities, and its inseparable accidents.

Averröes thinks that a type 2 proposition does not combine with a necessary major in a syllogism.[10] This is correct. A type 2 proposition has *per accidens* subject and predicate. But necessity-propositions have *per se* subjects. Therefore the combination in question lacks a genuine middle term. Averröes says that when Aristotle speaks of a necessity-premise in a mixed modal syllogism, he means a proposition of type 1, and when in the same context he speaks of an assertoric proposition he means a proposition of type 3. This is an original interpretive strategy, which has the interesting consequence that the necessity-conclusion in a mixed first-figure syllogism takes a different form from that of the necessity-premise. The conclusion turns out to have the same form as the assertoric minor premise. Both are type 3 propositions – *per accidens* necessities that are *per se* assertorics.

It appears then that Averröes is not putting forward a general division of either inclusions or of necessary inclusions. He is listing the types of inclusions that can figure in first-figure syllogisms. Type 1 propositions occur in uniform necessity-syllogisms, either as premise or as conclusion. They also occur in mixed necessity-syllogisms, as the major premise. Type 2 propositions – which, according to Averröes, are what Aristotle is primarily talking about in the *Analytics* – occur, presumably, in non-modal syllogisms. Type 3 propositions occur as the minor premise or the conclusion of mixed necessity-syllogisms. Type 4 propositions – which, Averröes says, do not occur in scientific syllogisms – do not occur in

[10] Averröes (1562) Q.4 ch.3, 84E.

necessity-syllogisms; though he seems to leave it open that they might occur in non-modal syllogisms. For the purposes of expounding Averröes's understanding of modal syllogistic, therefore, we may concentrate on types 1 and 3.

Type 1 propositions are necessity-inclusions of the form '$\Box\ B \rightarrow A$'. Analogous forms can be defined for the other *per se* necessity-propositions.

Definition 5.1 (La *per se*). 'Every *b* is *per se* necessarily *a*': $\Box\ B \rightarrow A$. There are three truth-conditions here: $\Box\ b \rightarrow a$, and $\Box\ b \rightarrow b^*$, and $\Box\ a \rightarrow a^*$.

Definition 5.2 (Le *per se*). 'No *b* is *per se* possibly *a*': $\Box\ B \mid A$. Again there are three truth-conditions: $\Box\ b \mid a$, and $\Box\ b \rightarrow b^*$, and $\Box\ a \rightarrow a^*$. In other words, universal *per se* necessities are *de dicto* necessities with *per se* terms. This being so, we can define particular *per se* necessities analogously.

Definition 5.3 (Li *per se*). 'Some *b* is *per se* necessarily *a*': $\Box\ b \smile a$ and $\Box\ b \rightarrow b^*$ and $\Box\ a \rightarrow a^*$.

Definition 5.4 (Lo *per se*). 'Some *b* is *per se* necessarily not *a*': $\Box\ b \nrightarrow a$ and $\Box\ b \rightarrow b^*$ and $\Box\ a \rightarrow a^*$.

Type 3 propositions are of the form '$\Box\ b \rightarrow A$' but not of the form '$\Box\ B \rightarrow A$'. They are *de dicto* necessities with *per accidens* subjects and *per se* predicates. Their truth-conditions are as follows.

Definition 5.5 (Xa *per se*). 'Every *b* is *per se a*': $\Diamond\ b \nrightarrow b^*$ and $\Box\ a \rightarrow a^*$ and $\Box\ b \rightarrow a$.

Definition 5.6 (Xe *per se*). 'No *b* is *per se a*': $\Diamond\ b \nrightarrow b^*$ and $\Box\ a \rightarrow a^*$ and $\Box\ b \mid a$.

Definition 5.7 (Xi *per se*). 'Some *b* is *per se a*': $\Diamond\ b \nrightarrow b^*$ and $\Box\ a \rightarrow a^*$ and $\Box\ b \smile a$.

Definition 5.8 (Xo *per se*). 'Some *b* is *per se* not *a*': $\Diamond\ b \nrightarrow b^*$ and $\Box\ a \rightarrow a^*$ and $\Box\ b \nrightarrow a$. These definitions assume that particular assertorics should stand to their universals in the same way that particular necessity-propositions stand to their universals. This is one way of answering Alexander's question, noted in Chapter 2,[11] about the meaning of particular *simpliciter* assertorics. It does, however, leave one question unanswered, namely the question what are the contradictories of these assertorics. This problem is not addressed by Averröes, and I will not broach it.

Per se assertoric propositions are not implied by the corresponding *per se* necessity-propositions, because the latter do not imply the *per accidens* status of the subject. In other words, the standard principles of modal subordination do not work on these readings. Averröes's position here is unorthodox, in so far as he defines types of proposition as mutually

[11] p.29 above.

exclusive, and eschews general definitions that embrace more than one propositional type. Had he taken the latter approach, he might have defined a general type of necessity-proposition of which both type 1 and type 3 propositions were species.

With this semantics we have the convertibility of *per se* apodeictic propositions. Consider the universal affirmative *per se* necessity-proposition stating that $\Box\, B \rightarrow A$. If this is true then there is a '*d*', namely '*b*' such that $\Box\, A \leftarrow D$ and $\Box\, D \rightarrow B$. The universal negative *per se* necessity-proposition stating that $\Box\, B \mid A$ implies that $\Box\, A \mid B$. *Per se* assertorics are not convertible because they do not imply that their predicates are *per accidens*.

All standard syllogisms are valid if transformed into uniform necessity-syllogisms with *per se* necessities. The reason is that under that transformation they are uniform *de dicto* necessity-syllogisms, governed by Rule 1.3; and the truth of the premises guarantees that the terms in the conclusion are *per se*. Averröes notes this result for the first figure, saying that if both premises are *per se* necessary then so is the conclusion.[12]

Theorem 5.1. The LLL moods are valid in Figure 1 in the *per se* sense.

Averröes does not deal in this work with syllogisms in the other figures. We can, however, infer some results on the basis of his principles.

[**Theorem 5.2.** The LLL moods are valid in Figure 2 in the *per se* sense.]

[**Theorem 5.3.** The LLL moods are valid in Figure 3 in the *per se* sense.]

These results can be proved in the same way as Theorem 5.1.

Assuming that we define M-propositions as contradictory to L-propositions of the opposite quantity and quality, and bearing in mind that universal necessity-propositions have three truth-conditions, we have the following definitions of *per se* possibility-propositions. In each case there are three disjuncts, each of which is a *de dicto* possibility-statement.

Definition 5.9 (Ma *per se*). 'Every *b* is possibly *a*': EITHER $\Diamond\, b \rightarrow a$ OR $\Diamond\, b \nrightarrow b^*$ OR $\Diamond\, a \nrightarrow a^*$.

Definition 5.10 (Me *per se*). 'No *b* is necessarily *a*': EITHER $\Diamond\, b \mid a$ OR $\Diamond\, b \nrightarrow b^*$ OR $\Diamond\, a \nrightarrow a^*$.

Definition 5.11 (Mi *per se*). 'Some *b* is possibly *a*': EITHER $\Diamond\, b \smile a$ OR $\Diamond\, b \nrightarrow b^*$ OR $\Diamond\, a \nrightarrow a^*$.

Definition 5.12 (Mo *per se*). 'Some *b* is not necessarily *a*': EITHER $\Diamond\, b \nrightarrow a$ OR $\Diamond\, b \nrightarrow b^*$ OR $\Diamond\, a \nrightarrow a^*$.

Given these definitions, Theorems 5.1 to 5.3 generate six corollaries by virtue of indirect reductions.

[**Theorem 5.1a.** The LMM moods are valid in Figure 2 in the *per se* sense.]

[**Theorem 5.1b.** The MLM moods are valid in Figure 3 in the *per se* sense.]

[12] Averröes (1562) Q.4 ch.3, 84F.

[Theorem 5.2a. The LMM moods are valid in Figure 1 in the *per se* sense.]
[Theorem 5.2b. The LMM moods are valid in Figure 3 in the *per se* sense.]
[Theorem 5.3a. The MLM moods are valid in Figure 1 in the *per se* sense.]
[Theorem 5.3b. The MLM moods are valid in Figure 2 in the *per se* sense.]

As a matter of interest, the MMM moods are invalid in Figure 1. On the basis of Definition 5.9 we can see that in order for Barbara MMM to be valid, its M-conclusion would need to be implied by each of the following assumptions:

1. $\Diamond b \rightarrow a$ and $\Diamond c \rightarrow b$
2. $\Diamond b \nrightarrow b^*$ and $\Diamond c \rightarrow b$
3. $\Diamond a \nrightarrow a^*$ and $\Diamond c \rightarrow b$
4. $\Diamond b \rightarrow a$ and $\Diamond c \nrightarrow c^*$
5. $\Diamond b \nrightarrow b^*$ and $\Diamond c \nrightarrow c^*$
6. $\Diamond a \nrightarrow a^*$ and $\Diamond c \nrightarrow c^*$
7. $\Diamond b \rightarrow a$ and $\Diamond b \nrightarrow b^*$
8. $\Diamond b \nrightarrow b^*$ and $\Diamond b \nrightarrow b^*$
9. $\Diamond a \nrightarrow a^*$ and $\Diamond b \nrightarrow b^*$.

In fact, it does not even follow from assumption 1. It is possible that everything running is a horse, and it is possible that every man is running; but it is not possible that every man is a horse, and it is the case that both 'man' and 'horse' are *per se*.

The MMM moods in Figure 3 are also invalid. For example, Disamis MMM is not valid, because its premises would be true and its conclusion false if the middle term were *per accidens*, and major and minor were mutually exclusive *per se* terms. A counter-example would be 'Something running is possibly a man, everything running is possibly a horse, so some horse is possibly a man'. Along with the MMM moods in Figure 3, it will be necessary to reject the LML moods in Figures 1 and 2.

In the *per se* sense, we can therefore attribute to Averröes the same L/M combinations that we attributed to Aristotle, as is shown in Table 5.1.

Table 5.1 Averröes's system: *per se* L/M combinations

Figure	*Combinations*
1	LLL, LMM, MLM
2	LLL, LMM, MLM
3	LLL, LMM, MLM

Averröes can be seen as having improved on Abélard's results (Table 3.6), by including the second-figure LLL moods, which Abélard had regarded as invalid. At the same time, he can be seen as having improved on Avicenna's results (Table 4.1), in so far as his semantics does not validate the MMM moods in Figures 1 and 3.

Let us turn to the mixed moods. According to Averröes the minor in a valid mixed first-figure necessity-syllogism can only be of type 3.[13] If both premises are of type 1, the syllogism is not mixed but pure, with two *per se* necessity-premises yielding a *per se* necessity-conclusion.[14] If the minor is type 1 and the major is type 2 or 4, nothing follows, because the subject of the major is *per accidens* and the predicate of the minor *per se*.[15] If the minor is of type 2, its predicate is *per accidens* and so the major must be of type 4, so that the conclusion cannot be a necessity-proposition. If the minor is of type 4 the major cannot be of type 1, because that would require the middle term to be both *per se* and *per accidens*.[16] Averröes says that if the major is necessary *per se*, and a *per se ut nunc* assertoric *per accidens* necessity is added, sometimes the conclusion is necessary, sometimes non-necessary:[17]

> An example where a necessity is concluded is when we say 'Everything white is now an animal, every animal is a body', which implies 'Everything white is a body', which is a necessity of the *per accidens* species. An example where a contingency is concluded is when we say 'Everything white is now a man, every man is sensitive', which implies 'Everything white is sensitive'. This, however, is only necessary for the times at which only white things are men.[18]

The only remaining possibility is that the minor is type 3. Averröes says that if the major is *per se* necessary and the minor *per accidens* necessary and *per se* assertoric, the conclusion will be *per accidens* necessary and assertoric *per se*.[19] Thus a syllogism that we have been categorizing as LXL in the first figure is not seen by Averröes as exhibiting that form. He views the conclusion in such a syllogism as having a different form from the major premise. He has two ways of describing the form of such a syllogism. According to one description, the syllogism contains propositions of three different forms – a *per se* necessity-major, a *per se* assertoric minor, and a *per* accidens necessity-conclusion. According to the second description, the syllogism contains propositions of two distinct

---

[13] Averröes (1562) Q.4 ch.3 84E.
[14] Averröes (1562) Q.4 ch.3 84F.
[15] Averröes (1562) Q.4 ch.3 84.
[16] Averröes (1562) Q.4 ch.3, 84.
[17] Averröes (1562) Q.4 ch.3, 84H.
[18] Averröes (1562) Q.4 ch.3, 84I.
[19] Averröes (1562) Q.4 ch.3, 84G.

forms; but it is the minor and the conclusion, not the major and the conclusion, that are of the same form, and that form can be characterized either as the form of *per se* assertorics or as that of *per accidens* necessity-propositions.

What this means is that Averröes views the Aristotelian LXL moods as exhibiting the form LXX, where 'X' stands for *per se* assertorics. Thus we have the following theorem.

Theorem 5.4. The LXX moods are valid in Figure 1 in the *per se* sense. Barbara: if $\Box\, B \rightarrow A$ and $\Diamond\, c \nrightarrow c^*$ and $\Box\, c \rightarrow B$ then $\Diamond\, c \nrightarrow c^*$ and $\Box\, c \rightarrow A$. Celarent: if $\Box\, B \mid A$ and $\Diamond\, c \nrightarrow c^*$ and $\Box\, c \rightarrow B$ then $\Diamond\, c \nrightarrow c^*$ and $\Box\, c \mid A$. Darii: if $\Box\, B \rightarrow A$ and $\Diamond\, c \nrightarrow c^*$ and $\Box\, c \smile B$, then $\Diamond\, c \nrightarrow c^*$ and $\Box\, c \smile A$. Ferio: if $\Box\, B \mid A$ and $\Diamond\, c \nrightarrow c^*$ and $\Box\, c \smile B$, then $\Diamond\, c \rightarrow c^*$ and $\Box\, c \nrightarrow A$.

Theorem 5.5. The XLL moods are valid in Figure 1 in the *per se* sense. Premises requiring that the subject of the major be *per accidens* and the predicate of the minor to be *per se* are mutually inconsistent, and so imply any conclusion.

What is not valid in Figure 1 is an inference from a *per se* necessity-major, and a *per se* assertoric minor, to a *per se* necessity-conclusion. Such premises, far from guaranteeing that the subject of the conclusion is *per se*, ensure that it is *per accidens*.

As a matter of interest, not noted by Averröes, we mention the following results for Figures 2 and 3.

[**Theorem 5.6.** The LXX moods are valid in Figure 2.] Cesare: if $\Box\, B \mid A$ and $\Diamond\, c \rightarrow c^*$ and $\Box\, c \rightarrow A$, then $\Diamond\, c \nrightarrow c^*$ and $\Box\, c \mid B$. Camestres: if $\Box\, B \rightarrow A$ and $\Diamond\, c \nrightarrow c^*$ and $\Box\, c \mid A$, then $\Diamond\, c \nrightarrow c^*$ and $\Box\, c \mid B$. Festino: if $\Box\, B \mid A$ and $\Diamond\, c \nrightarrow c^*$ and $\Box\, c \smile A$, then $\Diamond\, c \nrightarrow c^*$ and $\Box\, c \nrightarrow B$. Baroco: if $\Box\, B \rightarrow A$ and $\Box\, c \nrightarrow A$, then $\Box\, c \nrightarrow B$.

[**Theorem 5.7.** The XXL moods are valid in Figure 3.] Datisi: if $\Diamond\, c \nrightarrow c^*$ and $\Box\, \underline{c} \rightarrow A$ and $\Box\, \underline{c} \smile B$, then $\Box\, B \smile A$. Disamis: if $\Diamond\, c \nrightarrow c^*$ and $\Box\, \underline{c} \smile A$, and $\Box\, \underline{c} \rightarrow B$, then $\Box\, B \smile A$. Ferison: if $\Diamond\, c \nrightarrow c^*$ and $\Box\, c \mid A$ and $\Box\, c \smile B$, then $\Box\, B \nrightarrow A$. Bocardo: if $\Diamond\, c \nrightarrow c^*$ and $\Box\, c \nrightarrow A$, and $\Box\, c \rightarrow B$, then $\Box\, B \nrightarrow A$.

Averröes is clearly interested in cases where necessity-propositions have different senses (*per se* or *per accidens*) in the premises and the conclusion of an argument. This is interesting from a metaphysical point of view. A proposition can have a claim to metaphysical necessity, within the context of Aristotelian essentialism, only by virtue of standing in a relation to one of these *per se* necessary propositions. For example, the proposition 'Everything approaching is an animal', though not itself *per se* necessary, is necessary in a derivative (*per accidens*) way, provided that everything approaching is human, because in that case this proposition, being a logical consequence of the truth that everything approaching is human and the *per se* necessity 'Every human is an animal', inherits a derivative and secondary type of necessity.

From a logical point of view, we may wonder whether any inferences are valid from a *per se* assertoric premise and a necessity-premise to a necessity-conclusion of the same form as the necessity-premise. This does not happen with *per se* necessity-propositions. Notwithstanding the fact that Averröes does not talk in this way, we could construct a broad class of necessity-propositions embracing both *per se* and *per accidens* necessities. These propositions would have to have *per se* predicates, but their subjects could be either *per se* or *per accidens*. The characteristic forms would be '$\Box\, b \rightarrow A$', '$\Box\, b \mid A$', and so on. The LXL moods of Figure 1 would then be valid. Consider Barbara: if $\Diamond\, c \not\twoheadrightarrow c^*$ and $\Box\, c \rightarrow B$ and $\Box\, b \rightarrow A$, then $\Diamond\, \mathrm{c} \not\twoheadrightarrow c^*$ and $\Box\, c \rightarrow A$. The XLL moods of Figure 1 will also be valid, because the premises will be mutually inconsistent, the major requiring that the middle term be *per accidens*, the minor requiring that it be *per se*. So the construction of this broad class of necessities would not help Averröes achieve his aim of truth to Aristotle.

Summary

Averröes's late investigations into modal syllogistic do not propound an unambiguous theory. Nonetheless, two features of his thought appear to emerge. First there is his explicit deployment of essentialist notions in an attempt to clarify Aristotle's logic. He distinguishes a *per se* from a *per accidens* sense of modal propositions, making use of the *per se* sense when dealing with the LLL moods. This idea will find a following in thirteenth-century Latin writers, led by Robert Kilwardby. The *per se* in any domain is determined not just by the way things are, but by the way they are described (for example, to deny that the Chancellor *per se* is a member of the Board of Governors is to speak not simply of the human being who is Chancellor but rather of that person under the description 'Chancellor'). A *per se* necessity would therefore be one that depended not just on how things are but on how they are described. Averröes's use of the notion of *per se* necessity thus stands (along with Avicenna's ampliationist analysis) as an alternative to the Abélardian *de rebus* analysis of modals. It is more successful than the ampliationist approach in that it secures all the Aristotelian modal conversion-principles, not just those for Le- and Mi-propositions. By virtue of those principles, the pure necessity-moods in Figures 2 and 3, other than Baroco and Bocardo, reduce to the first figure, thus permitting a base consisting of the first-figure LLL moods along with Baroco and Bocardo LLL.

Second, Averröes thinks that in mixed syllogisms the conclusion expresses a different, and derivative, kind of necessity than does the modal premise. In other words, he sees the syllogism as marking a transition from a central strong sense of necessity to a derivative sense, rather than simply as a logical link between two necessity-propositions mediated by an assertoric. This way of seeing mixed necessity-syllogisms will be taken up in the twentieth century by Ulrich Nortmann, who proposes a 'rule of thumb' according to which, when an equivocal expression occurs in a premise as well as in the conclusion of a syllogism, it should be interpreted in its strong sense in the premise and in its weak sense in the conclusion.[20]

[20] Nortmann (1996) p.126.

Chapter 6

Kilwardby

The commentary, as a literary form, took a great many different shapes in the traditions stemming from Aristotle. In Alexander's hands, besides serving as an exposition of the text's literal meaning, it functioned as a repository of perceived difficulties – difficulties that the commentator felt no obligation to dispel. Abélard (at least in the introduction to his commentary on Chapter 12 of *De Interpretatione*) certainly strove to dissolve difficulties, but his main objective was to expound, rather than expose, a theory; for it was his own theory, not Aristotle's, that formed the focus of this introduction. With Robert Kilwardby (d.1279) the commentary takes on a new shape: there is a literal exposition punctuated by a series of *dubitationes* [puzzles] to which Robert offers carefully argued solutions, which hang together from puzzle to puzzle in ways that seem to the sympathetic reader to form an emerging theory; but the theory never fully emerges, and so the reader is ultimately disappointed.

Modal conversion: necessity-propositions

Abélard's rejection of Aristotle's laws of modal conversion must remain unsatisfying to an interpreter in search of the meaning of chapters 3 and 8 to 22 of book A of the *Prior Analytics*. The interpreter is left wondering why Aristotle accepted modal conversions, and naturally looks for ways of saving the laws of modal conversion. This is what Kilwardby does in his commentary.

According to Aristotle's doctrine of modal conversion, an affirmative necessity-proposition converts to a particular affirmative necessity-proposition – for example, if every man is necessarily an animal then some animal is necessarily a man.[1] While the inference may seem harmless in this instance, it is *prima facie* not valid in general, since, as Abélard saw, all that follows from the premise that some *b* is necessarily *a* is that something that is necessarily *a* is *b*. In order for us to be warranted in

[1] Aristotle (1989) A3, 25a32-33.

drawing the further conclusion that some *a* is necessarily *b*, we would need to be assured that every *b* is necessarily *b*.

Kilwardby discusses an apparent counter-example to Aristotle's laws of modal conversion. It is true that everything literate is necessarily a man, but it is not true that some man is necessarily literate. Literacy is found only in men, and necessarily so, but when it is found in any given man, it is so found contingently, not as a matter of necessity.

The term 'literate' [*grammaticus*] is not chosen by Kilwardby at random. As we noted in Chapter 1, this term – or its Greek equivalent '*grammatikòs*' – occurs in Aristotle's *Categories*, where it is given as an example of a denominative term, named after the abstract '*grammatiké*'.[2] Aristotle notes that grammatical knowledge, which is denoted by the abstract noun, cannot exist except in a soul.[3] He himself does not draw the consequence – which is obvious enough – that it is necessarily the case that whatever has grammatical knowledge is ensouled. This consequence appears in medieval examples as the proposition 'Whatever is literate is necessarily human'.

Kilwardby offers two solutions to the problem that this example generates for Aristotle's doctrine of modal conversion. The first solution turns on the fact that in a conversion one term occurs first as predicate then as subject. Kilwardby states that abstract terms are understood differently depending on whether they occur in subject- or predicate-position: in subject-position they stand for their *supposita*, in predicate-position for a quality or form. Thus, the term 'literate' stands for the literate in the proposition 'The literate are necessarily human', but it stands for literacy in the proposition 'Some human is necessarily literate'. It is clear, however, that this solution does not get to the heart of the matter, since the problem with 'literate' does not recur with all denominative terms. A denominative term denoting a differentia – such as 'rational' – poses no problem for modal conversion, since it is true both that the rational are necessarily human and that humans are necessarily rational. The salient feature of the literacy example is not that 'literate' is a denominative term but that literacy is an accident that is necessarily restricted to humankind.

Kilwardby's second solution recognizes this fact:

> These propositions, in which the name of an accident is subject, are not *per se* necessary but only *per accidens*.... Therefore, when Aristotle teaches that

---

[2] Aristotle (1963) 1a11-14.

[3] Aristotle (1963) 1a24-26.

> necessity-propositions convert, he only teaches that *per se* necessity-propositions convert....[4]

Lagerlund notes the resemblance between Kilwardby's language here and Averröes's terminology.[5] Like Averröes, Kilwardby draws a distinction between substantial (or essential) and accidental terms, counting among the former 'man', 'animal' and 'ass'; among the latter 'white', 'black' and 'literate';[6] and his 'substantial' terms seem to correspond exactly with Averröes's *per se* terms. Also like Averröes, Kilwardby uses the expression '*per se* necessity'. In Averröes (in the case of universal propositions) this refers to *de dicto* inclusions or exclusions with *per se* subject and predicate (Definitions 5.1 to 5.4); and it appears to have the same meaning in Kilwardby. However, unlike Averröes, Kilwardby does not claim to base the notion of a *per se* necessity on *Prior Analytics* A8. His commentary on that chapter contains no trace of Averröes's interpretation.

Given all this, Kilwardby is right to say that the Aristotelian laws of conversion hold for all *per se* necessity-propositions. If $\Box\, B \rightarrow A$ then for some '*d*' (namely '*b*') $\Box\, D \rightarrow A$ and $\Box\, D \rightarrow B$; and if $\Box\, B \mid A$ then $\Box\, A \mid B$. What is not true is that this is the only way of saving necessity-conversion. We can delineate a broad class of necessity-propositions, including *per accidens* necessities like 'The literate are necessarily human' as well as *per se* necessities, such that necessity-propositions in this broad sense are convertible. Necessity-propositions in this broad sense may have subjects that are not *per se*. It should also be allowed that the predicate not be *per se* (in order to allow for necessary accidents). We must also take into account Kilwardby's differential treatment of affirmative and negative necessity-propositions:

> ...in a negative necessity-proposition... the predicate is actually denied of all that is or that can be under the subject....[7]

> ...a universal affirmative necessity-proposition affirms the predicate of those things that are actually under the subject, and not of those that can be under the subject.[8]

---

[4] Kilwardby (1516) 7rb [*ad* A3, *dub*.4].
[5] Lagerlund (2000) p.32.
[6] Kilwardby (1516) 22ra [*ad* A14, *dub*.2].
[7] Kilwardby (1516) 27ra [*ad* A16, *dub*.5].
[8] Kilwardby (1516) 27ra [*ad* A16, *dub*.7].

Because of this feature of negative necessities, Kilwardby denies that the proposition 'Nothing white is black' expresses a necessity.[9] We can define such a broad sense as follows.

Definition 6.1 (La – broad sense). 'Every *b* is necessarily *a*': $\Box\, b \rightarrow a$.

Definition 6.2 (Le – broad sense). 'No *b* is possibly *a*': $\Box\, b^{\dagger} \mid a^{\dagger}$. These propositions are convertible.

Definition 6.3 (Li – broad sense). 'Some *b* is necessarily *a*': $\Box\, b \smile a$. These propositions are convertible.

Definition 6.4 (Lo – broad sense). 'Some *b* is not possibly *a*': $\Box\, b^{\dagger} \nrightarrow a^{\dagger}$.

The contradictories of these necessity-propositions are as follows.

Definition 6.5 (Ma). 'Every *b* is possibly *a*': $\Diamond\, b^{\dagger} \rightarrow a^{\dagger}$.

Definition 6.6 (Me). 'No *b* is necessarily *a*': $\Diamond\, b \mid a$.

Definition 6.7 (Mi). 'Some *b* is possibly *a*': $\Diamond\, b^{\dagger} \smile a^{\dagger}$.

Definition 6.8 (Mo). 'Some *b* is not necessarily *a*': $\Diamond\, b \nrightarrow a$.

Necessity-Syllogisms

L/M combinations

By confining necessity-propositions to the *per se* we can secure the validity of the pure necessity-syllogisms (Theorems 5.1 to 5.3). So much was evident to both Averröes and Kilwardby. What, perhaps, was not evident to them is that those syllogisms are still valid if necessity-propositions are taken in the broad sense.

Theorem 6.1. The broad-sense LLL moods are valid in Figure 1. Barbara: by Axiom 1.3, if $b \rightarrow a$ and $c \rightarrow b$ then $c \rightarrow a$; so, applying Rule 1.3, Barbara LLL is valid. Celarent: by Axiom 1.4, if $b^{\dagger} \mid a^{\dagger}$ and $c^{\dagger} \rightarrow b^{\dagger}$ then $c^{\dagger} \mid a^{\dagger}$; so if $\Box\, b^{\dagger} \mid a^{\dagger}$ and $\Box\, c^{\dagger} \rightarrow b^{\dagger}$ then $\Box\, c^{\dagger} \mid a^{\dagger}$. But if $\Box\, c \rightarrow b$ then $\Box\, c^{\dagger} \rightarrow b^{\dagger}$ (Axiom 1.14); whence follows the validity of Celarent LLL. Darii and Ferio can be proved similarly.

By indirect reduction this theorem has two corollaries.

[**Theorem 6.1a.** The broad-sense LMM moods are valid in Figure 2.]

[**Theorem 6.1b.** The broad-sense MLM moods are valid in Figure 3.]

Theorem 6.2. The broad-sense LLL moods are valid in Figure 2. Cesare: if $\Box\, b^{\dagger} \mid a^{\dagger}$ and $\Box\, c \rightarrow a$ then we may replace the second premise by '$\Box\, c^{\dagger} \rightarrow a^{\dagger}$' (Axiom 1.14), and, if $b^{\dagger} \mid a^{\dagger}$ and $c^{\dagger} \rightarrow a^{\dagger}$ then $c^{\dagger} \mid b^{\dagger}$ (Axiom 1.4); so by Rule 1.3 our replacement premises imply '$\Box\, c^{\dagger} \mid b^{\dagger}$'. Camestres: if $\Box\, b \rightarrow a$ and $\Box\, c^{\dagger} \mid a^{\dagger}$ then we may replace the major premise with '$\Box\, b^{\dagger} \rightarrow a^{\dagger}$'

[9] Kilwardby (1516) 17ra [*ad* A9, note].

(Axiom 1.14), and since the chain '$c^\dagger \mid a^\dagger \leftarrow b^\dagger$' simplifies to '$c^\dagger \mid b^\dagger$', if $\Box\ c^\dagger \mid a^\dagger$ and $\Box\ a^\dagger \leftarrow b^\dagger$ then $\Box\ c^\dagger \mid b^\dagger$. Festino: since the chain '$b^\dagger \mid a^\dagger \leftarrow a \smile c \rightarrow c^\dagger$' simplifies to '$c^\dagger \nrightarrow b^\dagger$' and its second and fourth links are axiomatic, we may argue '$b^\dagger \mid a^\dagger$ and $c \smile a$, so $c^\dagger \nrightarrow b^\dagger$', and thus '$\Box\ b^\dagger \mid a^\dagger$ and $\Box\ a \smile c$, so $\Box\ c^\dagger \nrightarrow b^\dagger$'. Baroco: if $\Box\ b \rightarrow a$ (so that $\Box\ b^\dagger \rightarrow a^\dagger$) and for some '$d$', $\Box\ d \rightarrow c^\dagger$ and $\Box\ d \mid a^\dagger$, then $\Box\ d \rightarrow c^\dagger$ and $\Box\ d \mid b^\dagger$.

With regard to Baroco LLL Kilwardby writes:

> And it is to be said, as some say, that there is no true exposition through singulars, but through the less universal. And those ought to be taken universally, thus: it is necessary for every man to be an animal, it is necessary for some white not to be an animal, therefore it is necessary for a white not to be a man. 'White' ought to be expounded, not for any designated sensible singular, but for some particular white such as snow. And then it ought to be taken universally. And a syllogism is made in the 2nd of the 2nd [sc. Camestres] thus: it is necessary for every man to be an animal, it is necessary for no snow to be an animal, therefore it is necessary for no snow to be a man; and so, since snow is some white, it is necessary for some white not to be a man.[10]

There are actually two syllogisms used in this proof, Camestres LLL (as Kilwardby notes) and Felapton LLL (arguing that since it is necessary for no snow to be a man and it is necessary that all snow is white, it is necessary for some white not to be a man).[11] In any case the modal syllogisms on which this expository proof rests are themselves reducible to the first figure by means of modal conversion.

Similar considerations apply to Bocardo LLL, which Kilwardby notes can be proved by an expository proof, relying on 'the 2nd of the 3rd' namely Felapton LLL. Again, a second syllogism will be involved in the proof, this time Barbara LLL.

Theorem 6.2 has two corollaries by virtue of indirect reduction.

[**Theorem 6.2a.** The broad-sense LMM moods are valid in Figure 1.]

[**Theorem 6.2b.** The broad-sense LMM moods are valid in Figure 3.]

Theorem 6.3. The broad-sense LLL moods are valid in Figure 3. This can be proved in the same way we proved Theorem 6.1.

Hence come two corollaries.

[**Theorem 6.3a.** The broad-sense MLM moods are valid in Figure 1.]

[**Theorem 6.3b.** The broad-sense MLM moods are valid in Figure 2.]

---

[10] Kilwardby (1516) 16ra [*ad* A8, *dub*.8]. It is striking that Kilwardby uses the very same terms to illustrate the validity of Baroco LLL as Aristotle had used (at *Prior Analytics* A10, 31a15-17) to demonstrate the invalidity of Baroco XLL.

[11] See Thom (1996) p.50.

Taking into account the broad-sense LMM and MLM syllogisms that are equivalent to LLL syllogisms by indirect reduction, we can attribute to Kilwardby the syllogisms listed in Table 6.1 (equivalently, Table 1.3).

Table 6.1 Kilwardby's broad-sense system: L/M combinations

Figure	*Combinations*
1	LLL, LMM, MLM
2	LLL, LMM, MLM
3	LLL, LMM, MLM

L/X/M combinations

Let us turn to mixed necessity-moods. Kilwardby is aware of apparent counter-examples to Barbara LXL, like 'Every man is necessarily an animal, everything white is a man, so everything white is necessarily an animal'.[12] He takes the conclusion here to be false and the major premise true; and he admits that we could suppose the minor to be true. However, he considers the counter-example to be irrelevant because the minor premise in Barbara LXL should be *simpliciter*, rather than being true merely *ut nunc* or accidentally. He explains that a *simpliciter* assertoric is a necessary proposition, even though it does not contain an express mode of necessity.[13] Kilwardby states that 'Socrates is a man' and 'This man is a man' are not *simpliciter* assertorics. We observe that these propositions lack the necessity that is required of *simpliciter* assertorics; they are necessary only on the condition that Socrates, or this man, exists. Kilwardby says that they are *ut nunc* predications. The proposition 'White Socrates is a man' is not *simpliciter*, but is a *per accidens* predication. This proposition is not unconditionally necessary, but is necessary only on the condition that Socrates exists and is white. Kilwardby's examples of *simpliciter* assertorics include 'Every animal is capable of sensation' and 'Some man is a man'.[14] (This second example makes a metaphysical assumption, namely that a natural kind-term such a 'man' is necessarily non-null.) He also admits as *simpliciter* assertorics the propositions

---

[12] Kilwardby (1516) 16vb [*ad* A9, *dub*.6]. Compare Alexander's counter-examples, p.28 above.

[13] Kilwardby (1516) 16va [*ad* A9, *dub*.2].

[14] Kilwardby (1516) 16vb-17ra [*ad* A9, *dub*.6].

'Whatever is white is coloured' and 'Whatever is walking is moving'.[15] These last examples make it seem that a sufficient condition for an assertoric to be *simpliciter* is that it be necessary; and Kilwardby does say that a *simpliciter* assertoric 'perhaps does not differ in fact from a necessary proposition, even if it differs in expression'.[16] With this in mind, we may perhaps define *simpliciter* assertorics as follows.

Definition 6.9 (Xa – *simpliciter*). 'Every b is a': $\Box\, b \rightarrow a$.

Definition 6.10 (Xe – *simpliciter*). 'No b is a': $\Box\, b^{\dagger} \mid a^{\dagger}$.

Definition 6.11 (Xi – *simpliciter*). 'Some b is a': $\Box\, b \smile a$.

Definition 6.12 (Xo – *simpliciter*). 'Some b is not a': $\Box\, b^{\dagger} \twoheadrightarrow\!\!\!\!\!\!/\;\; a^{\dagger}$.

Note that the contradictories of these *simpliciter* assertorics are possibilities as defined in Definitions 6.5 to 6.8. Given these definitions, we can enunciate several theorems.

Theorem 6.4. The LXL moods with broad necessities and *simpliciter* assertorics are valid in Figure 1. This is equivalent to Theorem 6.1.

By indirect reduction we have a corollary.

[**Theorem 6.4a.** The MXM moods with *simpliciter* assertorics are valid in Figure 3.][17]

Kilwardby considers the question whether XLL moods are valid in Figure 1, and comes to the (Aristotelian) conclusion that they are not, on the ground that in the first figure certain features of the conclusion mirror those of the major premise, specifically in relation to the major term, which is predicate in the conclusion and in that premise. The features in question are quality and modality.[18] By our semantics, however, these moods are valid.

[**Theorem 6.5.** The XLL moods in Figure 1 with broad necessities and *simpliciter* assertorics are valid.] This is again equivalent to Theorem 6.1.

Corollary:

[**Theorem 6.5a.** The XMM moods with *simpliciter* assertorics in Figure 2 are valid.]

Theorem 6.6. The LXL moods with broad necessities and *simpliciter* assertorics in Figure 2 are valid. This result is equivalent to Theorem 6.2.

---

[15] Kilwardby (1516) 24vb [*ad* A15, *dub.*6]. This last proposition had been categorized by Averröes as a *simpliciter* assertoric lacking 'any specific time'. See p.84 above.

[16] Kilwardby (1516) 21ra [*ad* A13, *dub.*10].

[17] Bearing in mind that the contradictory of a *simpliciter* assertoric is a possibility-proposition, we can also deduce Theorem 6.1a from Theorem 6.4; but since it has been proved already I do not mention it again here.

[18] Kilwardby (1516) 16vb [*ad* A9, *dub.*4].

Corollary:
[**Theorem 6.6a.** The XMM moods with *simpliciter* assertorics in Figure 3 are valid.].
[**Theorem 6.7.** The XLL moods with broad necessities and *simpliciter* assertorics in Figure 2 are valid.] This is equivalent to Theorem 6.3. Kilwardby denies that Baroco XLL is valid, at the same time noting that its validity appears to be provable in two ways – either from the validity of a Barbara with assertoric major, contingency-minor and contingency-conclusion, or by an expository syllogism in Camestres XLL.[19] (As we saw in Chapter 2,[20] this second derivation is as old as Alexander.) He reminds us, however, that the assertoric major premises in those enabling moods may be *ut nunc*, and that if they are then these purportedly enabling moods are not valid. This is correct. What is equally correct (and which he does not point out) is that if the majors in the enabling moods are *simpliciter* assertorics then those moods are valid, and therefore so is Baroco XLL under the same assumption about its assertoric major premise. Kilwardby's statements are all correct: Baroco XLL is, *sans phrase*, invalid; and its validity cannot be proved *sans phrase* by the proposed methods. However, he omits something that is equally correct, namely that with the addition of a suitable assumption – an assumption that he himself makes in many other cases – Baroco XLL is valid and can be proved by the proposed methods. This partial coverage of the truth, a truth that must have been known to him,[21] can be explained by supposing that he saw his primary task as expounding Aristotle's text with a view to preserving as much as possible of it as true; he did not see his primary task as developing a good modal theory.
Corollary:
[**Theorem 6.7a.** The XMM moods with *simpliciter* assertorics in Figure 1 are valid.]
Theorem 6.8. The LXL moods with broad necessities and *simpliciter* assertorics are valid in Figure 3. This is equivalent to Theorem 6.3. Kilwardby's treatment of Bocardo LXL is similar to his discussion of Baroco XLL. In discussing Aristotle's apparent reliance on this mood in proving the validity of Barbara XQM, Kilwardby says that this mood is 'useless simply speaking, because it is useless if the assertoric is *ut nunc*'.[22]

---

[19] Kilwardby (1516) 17vb [*ad* A10, *dub*.4].
[20] See above p.29.
[21] He notes explicitly that Aristotle uses *ut nunc* assertorics in counter-examples to invalid moods: Kilwardby (1516) 16vb [*ad* A9, *dub*.5].
[22] Kilwardby (1516) 24va [*ad* A15, *dub*.6].

[Theorem 6.8a. The MXM moods with *simpliciter* assertorics are valid in Figure 1.]
Theorem 6.9. The XLL moods with broad necessities and *simpliciter* assertorics in Figure 3 are valid. This is equivalent to Theorem 6.3.
[Theorem 6.9a. The MXM moods with *simpliciter* assertorics in Figure 2 are valid.]

Table 6.2 Kilwardby's broad-sense *simpliciter* system: L/X/M combinations

Figure	*Combinations*
1	LXL, XLL, XMM, MXM
2	LXL, XLL, XMM, MXM
3	LXL, XLL, XMM, MXM

This is a maximal system, containing all LXL and all XLL moods in all figures.

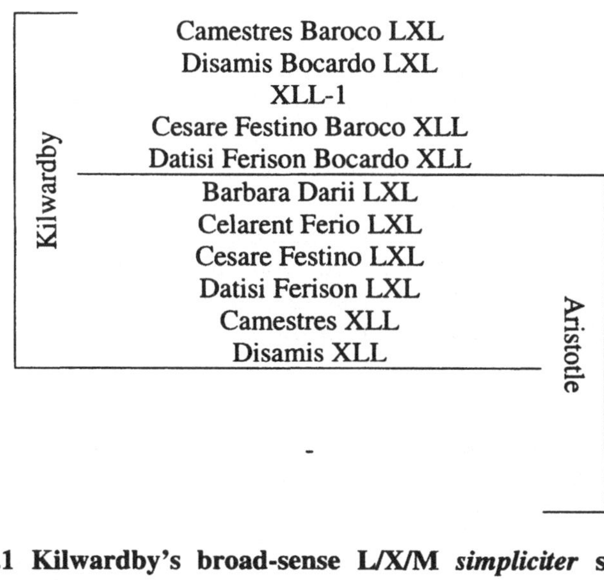

Figure 6.1 Kilwardby's broad-sense L/X/M *simpliciter* system and Aristotle's system

The relations with Aristotle's L/X/M system are displayed in Figure 6.1. In brief, Kilwardby's system includes Aristotle's and much else besides.

Contingency-syllogisms

Kilwardby follows Alexander's interpretive strategy whereby, in the case of pure contingency-syllogisms, both conclusion and premises are to be read as being ampliated, but in the case of mixed contingency-syllogisms this is not the case.

> For 'Every *b* may or may not be *a*' has a double meaning, namely either 'Everything that is *b* may or may not be *a*', or 'Everything that may or may not be *b* may or may not be *a*'.... It is this sense – 'For everything for which *b* is contingent, *a* is contingent' – that is taken in the generation of uniform contingency-syllogisms; however, it is the other sense that is taken in the mixtures of contingency- and assertoric or necessity-propositions.[23]

Clearly then we will need different semantic approaches for pure and for mixed contingency-syllogisms.

Q/Q combinations

We might therefore propose the following definitions in the context of pure contingency-syllogisms:
Definition 6.13 (Qa, Qe – ampliated). 'Every *b* may or may not be *a*': $b^{\ddagger} \rightarrow a^{\ddagger}$.
Definition 6.14 (Qi, Qo – ampliated). 'Some *b* may or may not be *a*': $b^{\ddagger} \smile a^{\ddagger}$.

Influenced perhaps by Averröes's idea that propositions can be classified according to the types of terms they contain, Kilwardby states that only certain types of terms can occur in the premises of uniform contingency-syllogisms if those premises are to be true:

> ...in general with uniform contingency-syllogisms, whenever the major's subject is an accidental term and its predicate a substantial term, the major is false, because the sense is 'Everything that is contingently *b* etc.'. Now, when such a proposition is taken as major (e.g. 'Everything that is white is contingently an animal'), this is false, because it is taken under this meaning that everything that is contingently white is contingently an animal, which is false.... Hence the middle term in this conjugation should not be an accidental term when the major term... is essential.[24]

[23] Kilwardby (1516) 19vb [*ad* A13].
[24] Kilwardby (1516) 22ra [*ad* A14, *dub.*2].

Why should the proposition 'Everything that is contingently white is contingently an animal' be false? For the following reason. There are terms – Kilwardby mentions 'wood' and 'stone' – for which being white is contingent but being an animal is impossible.[25]

Theorem 6.10. The QQQ moods are valid in Figure 1. Barbara (Celarent): if $b^‡ \rightarrow a^‡$ and $c^‡ \rightarrow b^‡$ then $c^‡ \rightarrow b^‡ \rightarrow a^‡$, and thus $c^‡ \rightarrow a^‡$. Darii (Ferio): if $b^‡ \rightarrow a^‡$ and $c^‡ \leftarrow d \rightarrow b^‡$ then $c^‡ \leftarrow d \rightarrow b^‡ \rightarrow a^‡$, and thus $c^‡ \leftarrow d \rightarrow a^‡$.

Theorem 6.11. The QQQ moods are valid in Figure 3. Datisi (Ferison): if $c^‡ \rightarrow a^‡$ and $c^‡ \leftarrow d \rightarrow b^‡$ then $b^‡ \leftarrow d \rightarrow c^‡ \rightarrow a^‡$ and so $b^‡ \leftarrow d \rightarrow a^‡$. Disamis (Bocardo): if $c^‡ \leftarrow d \rightarrow a^‡$ and $c^‡ \rightarrow b^‡$ then $b^‡ \leftarrow c^‡ \leftarrow d \rightarrow a^‡$ and so $b^‡ \leftarrow d \rightarrow a^‡$.

Particular contingency-propositions clearly are convertible, given the symmetry of '$b^‡ \leftarrow d \rightarrow a^‡$'. Datisi (Ferison) QQQ is therefore reducible in the Aristotelian fashion to Darii (Ferio) QQQ, as is Disamis (Bocardo) QQQ.

Kilwardby's results agree with Aristotle's (Table 1.5).

Table 6.3 Kilwardby's system: ampliated Q/Q combinations

Figure	*Combinations*
1	QQQ
2	-
3	QQQ

X/Q combinations

According to Kilwardby the minor premise in Barbara QXQ should be assertoric *simpliciter*.[26] He finds Aristotelian support for this requirement where Aristotle's text says that *c* should be 'under' *b*.[27] (In our own time, Rini has taken a similar view of this text.[28]) He thinks that X/Q combinations are valid when the assertoric minor is *simpliciter* but not otherwise.

How then are we to understand the contingency-propositions that occur in these syllogisms? Kilwardby tells us that these propositions are

[25] The text says 'Even if nothing but white things were animals, still becoming white is contingent for many things, for which being an animal is *possible*, such as wood and stone'; but it should read 'for which being an animal is *impossible*'.

[26] For this requirement, see Kilwardby (1516) 21rb [*ad* A13, *dub*.12].

[27] Aristotle (1989) A15, 33b33-36.

[28] Rini (2000).

unampliated.[29] He also believes that at least some propositions expressing contingencies hold necessarily:

> And it must be said that the act of greying can be said of the change towards greyness, as much as of the completion of that change. If it is said of the change towards greyness, a man is greying always and of necessity, so long as he exists. For greyness is produced by the incorporation of phlegm into the upper part of the head, as a result of the diminution of natural heat; and this diminution and incorporation exists always and continuously.[30]

Leaving aside the outdated physiology, the underlying idea here is that there is something about the human condition that sets up a process leading to greying, and that because of this it is necessary that humans have a tendency to go grey. This type of ('natural') contingency-proposition is defined in definitions 6.15 and 6.16.

Definition 6.15 (Qa, Qe – 'natural'). 'Every *b* may or may not be *a*': $\Box\, b \rightarrow a^{\ddagger}$.

Definition 6.16 (Qi, Qo – 'natural'). 'Some *b* may or may not be *a*': $\Box\, b \smile a^{\ddagger}$.

Theorem 6.12. The QXQ moods with 'natural' contingencies are valid in Figure 1 provided that assertoric propositions are understood *simpliciter*. Barbara (Celarent): if $b \rightarrow a^{\ddagger}$ and $c \rightarrow b$ then $c \rightarrow a^{\ddagger}$, therefore if $\Box\, b \rightarrow a^{\ddagger}$ and $\Box\, c \rightarrow b$ then $\Box\, c \rightarrow a^{\ddagger}$. Darii (Ferio): if $b \rightarrow a^{\ddagger}$ and $c \smile b$ then $c \smile a^{\ddagger}$, therefore if $\Box\, b \rightarrow a^{\ddagger}$ and $\Box\, c \smile b$ then $\Box\, c \smile a^{\ddagger}$.

On these definitions, contingency-propositions are not convertible, and so the third-figure QXQ moods will not be reducible in the Aristotelian manner to the first figure. Nevertheless these syllogisms are valid.

Theorem 6.13. The third-figure QXQ moods are valid with 'natural' contingencies and *simpliciter* assertorics. Datisi (Ferison): if $c \rightarrow a^{\ddagger}$ and $c \smile b$ then $b \smile a^{\ddagger}$, therefore if $\Box\, c \rightarrow a^{\ddagger}$ and $\Box\, c \smile b$ then $\Box\, b \smile a^{\ddagger}$. Disamis (Bocardo): if $c \smile a^{\ddagger}$ and $c \rightarrow b$ then $b \smile a^{\ddagger}$, therefore if $\Box\, c \smile a^{\ddagger}$ and $\Box\, c \rightarrow b$, then $\Box\, b \smile a^{\ddagger}$. Disamis XQQ, however, is not valid. If $\Box\, c \smile a$ and $\Box\, c \rightarrow b^{\ddagger}$ then we may infer that $\Box\, b^{\dagger} \smile a$, but not that $\Box\, b \smile a^{\ddagger}$. Necessarily some man is an animal and necessarily every man is contingently greying, but it is not the case that something that is greying is contingently an animal.

We summarize Kilwardby's X/Q contingency-syllogistic in Table 6.4. This system includes all of Aristotle's X/Q moods with the exception of

[29] Kilwardby (1516) 19vb [*ad* A13].

[30] Kilwardby (1516) 20va-b [*ad* A13, *dub.*8].

Disamis XQQ, whose proof in Aristotle depends on the convertibility of particular affirmative contingency-propositions.

Table 6.4 Kilwardby's 'natural' contingency *simpliciter* system: X/Q combinations

Figure	*Combinations*
1	QXQ
2	-
3	QXQ

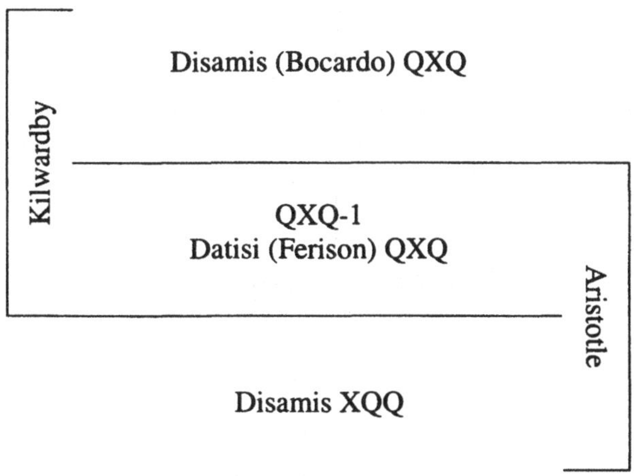

Figure 6.2 Kilwardby's 'natural' contingency X/Q *simpliciter* system and Aristotle's system

L/Q combinations

The 'natural' sense of contingency-propositions can also be used to validate Aristotle's L/Q combinations. Indeed, these moods are equivalent to the X/Q combinations, given that the truth-conditions for *simpliciter* assertorics are the same as those for necessity-propositions. Table 1.7 lists all the L/Q combinations that Aristotle takes to yield a Q-conclusion.

Theorem 6.14. The QLQ moods are valid in Figure 1. This is equivalent to Theorem 6.12.

Celarent LQL (which Alexander thought was valid)[31] is not valid on the present semantics. If $\Box\, b^{\dagger} \mid a^{\dagger}$ and $\Box\, c \rightarrow b^{\ddagger}$ then we may infer that $\Box\, c \mid a^{\dagger}$ but not that $\Box\, c^{\dagger} \mid a^{\dagger}$.

Theorem 6.15. The QLQ moods are valid in Figure 3. This is equivalent to Theorem 6.13.

Table 6.5 Kilwardby's system: 'natural' contingency L/Q combinations

Figure	*Combinations*
1	QLQ
2	-
3	QLQ

Disamis LQQ is invalid: if $\Box\, c \smile a^{*}$ and $\Box\, c \rightarrow b^{\ddagger}$ then we may infer that $\Box\, b^{\dagger} \smile a^{*}$, but not that $\Box\, b \smile a^{\ddagger}$. The failure of this mood is attributable to the failure of conversion for contingency-propositions as we have defined them. These results are shown in Table 6.5.

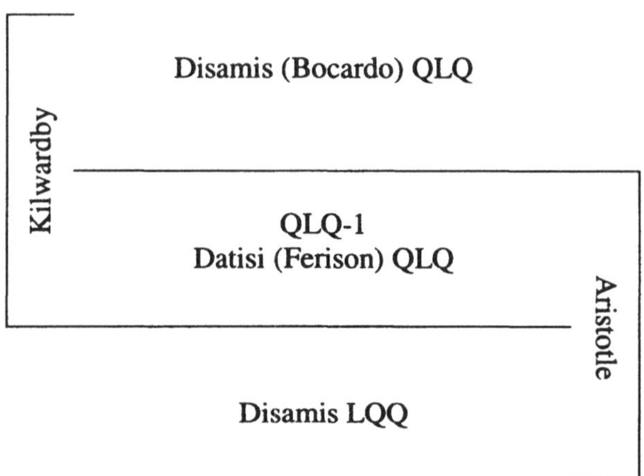

Figure 6.3 Kilwardby's 'natural' contingency L/Q system and Aristotle's system

[31] See above p.31.

L/X/Q combinations

Concerning the contingency-premise in these moods, Kilwardby says:

> In assertoric and necessity-[syllogisms] the minor needs to be taken as an affirmation of actual, not potential, inherence. But when a contingency is taken under a necessity, the minor is taken according to fitness for inherence [*secundum inherentiam aptitutinalem*].[32]

It appears that by 'fitness for inherence' Kilwardby means to invoke the 'natural' reading of contingency-propositions. This emerges in his discussion of the inference 'Of necessity no stone is a man, it may or may not be that everything moving is a stone, so nothing moving is a man', which appears to be a counter-example to Celarent LQX. Having first suggested that the difficulty could be solved by taking the contingency-premise as unampliated, Kilwardby goes on to state his preferred solution:

> It is better to say that [in the counter-example] contingency is taken in the contingent minor as the indefinite contingent, whereas it should be taken for natural contingency, as required by the necessity-major.[33]

It may or may not be that everything moving is a stone. There is nothing in the nature of movement or of stones that inclines things more one way than the other. This is the typical situation with 'indefinite' contingencies. With natural contingencies, on the other hand, such as the greying of humans with age, there is something in their nature that inclines matters in one direction.

Theorem 6.16. Celarent and Ferio LQX with 'natural' contingencies are valid. Celarent: since the chain '$c \rightarrow b^{\ddagger} \rightarrow b^{\dagger} \mid a^{\dagger} \leftarrow a$' simplifies (by Axioms 1.3 and 1.4) to '$c \mid a$', and the second and fourth links are axiomatic (Axioms 1.9 and 1.8), we can argue from '$b^{\dagger} \mid a^{\dagger}$' and '$c \rightarrow b^{\ddagger}$' to '$c \mid a$' and thus from '$\Box\, b^{\dagger} \mid a^{\dagger}$' and '$\Box\, c \rightarrow b^{\ddagger}$' to '$\Box\, c \mid a$' (by Rule 1.3) and thence to '$c \mid a$'. Ferio: since the chain '$c \smile b^{\ddagger} \rightarrow b^{\dagger} \mid a^{\dagger} \leftarrow a$' simplifies to '$c \nrightarrow a$', and the second and fourth links are axiomatic, we may argue '$\Box\, b^{\dagger} \mid a^{\dagger}$, $\Box\, c \smile b^{\ddagger}$, so $\Box\, c \nrightarrow a$' and thence to '$c \nrightarrow a$'.

It is noteworthy that Aristotle sees a linkage between the L/X/Q and L/X/M layers, proving the validity of Celarent and Ferio LQX on the basis of Ferio and Celarent LXL.[34] However, seeing that Ferio and Celarent LXL are valid for Kilwardby only with *simpliciter* assertorics, in order for the

---

[32] Kilwardby (1516) 26vb [*ad* A16, *dub*.2].

[33] Kilwardby (1516) 27 ra-b [*ad* A16, *dub*.5].

[34] See Thom (1996) p.58.

Aristotelian proofs to work those *simpliciter* premises would need to contradict the assertoric conclusions of Celarent and Ferio LQX. These conclusions, as we have just seen, are ordinary assertorics. Now, the contradictory of a *simpliciter* assertoric is not an ordinary assertoric, but the corresponding possibility-proposition. So within Kilwardby's system the Aristotelian proofs do not establish the validity of Celarent and Ferio LQX but only that of Celarent and Ferio LQM, and the Aristotelian linkage between the two layers is severed.

Barbara and Darii LQX are invalid. Their invalidity results from the difference that Kilwardby stipulates between affirmative and negative necessity-propositions. Kilwardby cites the counter-example 'Every literate being necessarily has knowledge, every man may or may not be literate, so every man has knowledge'.[35]

Theorem 6.17. Cesare and Festino LQX with 'natural' contingencies are valid. Cesare: since the chain '$c \rightarrow a^{\ddagger} \rightarrow a^{\dagger} \mid b^{\dagger} \leftarrow b$' simplifies to '$c \mid b$', and the second and fourth links are axiomatic (Axioms 1.9 and 1.8), we can argue from '$b^{\dagger} \mid a^{\dagger}$' and '$c \rightarrow a^{\ddagger}$' to '$c \mid b$' and thus from '$\Box\, b^{\dagger} \mid a^{\dagger}$' and '$\Box\, c \rightarrow a^{\ddagger}$' to '$\Box\, c \mid b$' and thence to '$c \mid b$'. Festino: since the chain '$c \smile a^{\ddagger} \rightarrow a^{\dagger} \mid b^{\dagger} \leftarrow b$' simplifies to '$c \nrightarrow b$' and the second and fourth links are axiomatic, we may argue '$\Box\, c \smile a^{\ddagger}$ and $\Box\, b^{\dagger} \mid a^{\dagger}$, so $\Box\, c \nrightarrow b$, so $c \nrightarrow b$'.

Camestres LQX is invalid: it is necessary that everything walking is moving, and it is necessary that every animal is contingently moving, but it does not follow that no animal is walking. Baroco too is invalid, as can be shown using the same terms.

Ferison LQX with 'natural' contingency-minor is invalid: it is necessary that men not be horses, and it is necessary that some man is potentially a musician, but it does not follow that some musician is not a horse (because there may be no musicians). The Aristotelian proof of Ferison LQX is not available, since it depends on the convertibility of contingency-propositions.[36] The other third-figure LQX moods are equally invalid. Datisi: all of the literate are necessarily human, and it may or may not be that some of literate are reading, but it does not follow that someone actually reading is human. Disamis: some of the literate are necessarily human, and all of the literate are contingently reading, but it does not follow that someone actually reading is human. Bocardo: some animal cannot be human, and every animal is contingently moving, but it does not follow that something actually moving is not human.

---

[35] Kilwardby (1516) 27va [*ad* A16, *dub*.8].

[36] See Thom (1996) p.63.

Theorem 6.18. Camestres and Baroco QLX with 'natural' contingencies are valid. Camestres: since the chain '$c \rightarrow c^{\dagger} \mid a^{\dagger} \leftarrow a^{\ddagger} \leftarrow b$' simplifies to '$c \mid b$', and the first and third links are axiomatic, we may argue from '$c^{\dagger} \mid a^{\dagger}$' and '$b \rightarrow a^{\ddagger}$' to '$c \mid b$', and hence from '$\Box\, c^{\dagger} \mid a^{\dagger}$' and '$\Box\, b \rightarrow a^{\ddagger}$' to '$\Box\, c \mid b$', and thence to '$c \mid b$'. Baroco: since the chain '$c \rightarrow c^{\dagger} \nrightarrow a^{\dagger} \leftarrow a^{\ddagger} \leftarrow b$' simplifies to '$c \nrightarrow b$', and the first and third links are axiomatic, we may argue from '$c^{\dagger} \nrightarrow a^{\dagger}$' and '$b \rightarrow a^{\ddagger}$' to '$c \nrightarrow b$', and hence from '$\Box\, c^{\dagger} \nrightarrow a^{\dagger}$' and '$\Box\, b \rightarrow a^{\ddagger}$' to '$\Box\, c \nrightarrow b$', and thence to '$c \nrightarrow b$'. Kilwardby considers a purported reduction of Baroco QLX to Barbara QXM; he replies that, since the minor in Barbara QXM must be *simpliciter* (Theorem 6.33), this proof does not justify an ordinary assertoric conclusion for Baroco QLX.[37]

Cesare and Festino QLX are invalid, as may be shown by adapting our counter-example to Camestres LQX.

These results are shown in Table 6.6 and compared with the Aristotelian results in Figure 6.4.

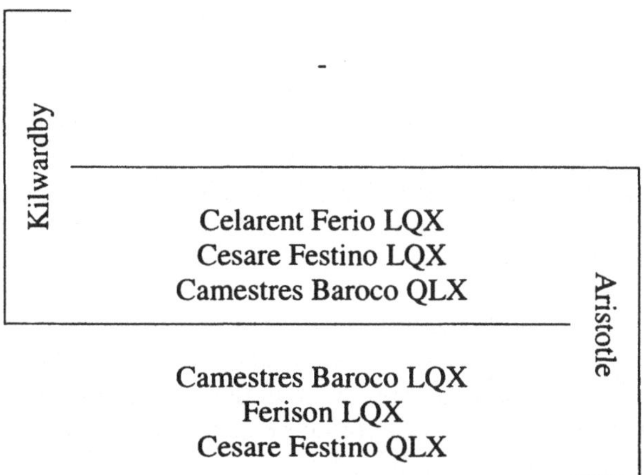

Figure 6.4 Kilwardby's 'natural' contingency L/X/Q system and Aristotle's system

[37] Kilwardby (1516) 31rb [*ad* A19, *dub*.11].

Table 6.6 Kilwardby's 'natural' contingency system: L/X/Q combinations

Figure	*Combinations*
1	Celarent Ferio LQX
2	Cesare Festino LQX, Camestres Baroco QLX
3	-

L/Q/M combinations

Just as Aristotle links the L/X/Q and L/X/M layers by reducing syllogisms in the former to syllogisms in the latter, so he links the L/Q/M and L/M layers, for example, by reducing first-figure LQM moods to LLL moods in the second figure. And in this case, there is no reason why Kilwardby cannot accept the linkage. All he needs is a way of understanding contingency-propositions that allows them to imply the corresponding possibility-propositions. This condition is satisfied by the following definitions.

Definition 6.17 (Qa, Qe – 'indefinite'). 'Every *b* may or may not be *a*': $\Diamond\ b^{\dagger} \rightarrow a^{\dagger}$ and $\Diamond\ b \mid a$. This is the conjunction of the corresponding Ma- and Me-propositions (Definitions 6.5 and 6.6).

Definition 6.18 (Qi, Qo – 'indefinite'). 'Some *b* may or may not be *a*': $\Diamond\ b \smile a$ and $\Diamond\ b \nrightarrow a$. The second conjunct here is the corresponding Mo-proposition (Definition 6.8), and the first conjunct implies the corresponding Mi-proposition (Definition 6.7).

The label 'indefinite' is appropriate here because, while these definitions are not suited to natural contingencies like the greying of humans, they are suited to indefinite contingencies such as the occurrence of thunder while someone is walking. There is nothing necessary about such conjunctions; their occurrence is purely contingent.

All LQM and QLM moods will be valid if we read contingency-propositions in this 'indefinite' sense. This is clear from the fact (Table 6.1) that all the LMM and MLM moods are valid in all figures, and that 'indefinite' contingencies imply the corresponding possibilities. This, therefore, is an appropriate reading – seeing that Aristotle accepts all the LQM and QLM moods (Table 1.8).

Theorem 6.19. The 'indefinite' LQM moods are valid in Figure 1.

First-figure LQM moods are reducible, in the Aristotelian manner, to second-figure LLL moods.[38]

Theorem 6.20. The 'indefinite' LQM moods are valid in Figure 2.

---

[38] See Thom (1996) p.69.

Aristotle reduces Cesare and Festino LQM, via the conversion of universal negative necessity-propositions, to first-figure LQM moods;[39] and these proofs are available to Kilwardby.

Theorem 6.21. The 'indefinite' LQM moods are valid in Figure 3.

Aristotle reduces Darapti, Felapton and Datisi LQM, by conversion of particular affirmative contingency-propositions, to first-figure LQM moods;[40] and these proofs are available to Kilwardby. However, it is also possible to reduce all third-figure LQM moods indirectly to first-figure LQM moods.[41]

Theorem 6.22. The 'indefinite' QLM moods are valid in Figure 1.

Aristotle reduces Ferio QLM indirectly to Datisi LLL,[42] and this method can be used to validate all the first-figure QLM moods.

Theorem 6.23. The 'indefinite' QLM moods are valid in Figure 2.

The second-figure QLM moods all reduce indirectly to the QLM moods in Figure 1.[43]

Theorem 6.24. The 'indefinite' QLM moods are valid in Figure 3.

Aristotle reduces Bocardo QLM indirectly to Barbara LLL,[44] and this method can be generalized to all the third-figure QLM moods.

Table 6.7 Kilwardby's 'indefinite' contingency system: L/Q/M combinations

Figure	*Combinations*
1	LQM, QLM
2	LQM, QLM
3	LQM, QLM

X/Q/M combinations

According to Aristotle, the assertoric propositions in the XQM moods have to be *simpliciter*. How then are we to read the modal propositions in these moods? We find a clue in Aristotle's proof of the validity of Barbara XQM with *simpliciter* assertoric. The proof has two parts. In the first part, Barbara XQM is reduced to a Bocardo with necessity-major, contingency-

[39] See Thom (1996) p.71.
[40] See Thom (1996) p.73.
[41] See Thom (1996) p.77.
[42] See Thom (1996) p.70.
[43] See Thom (1996) p.73. Kilwardby (1516) 31rb [*ad* A19, *dub*.12] points out that Baroco QLM can be reduced to Barbara QLQ.
[44] See Thom (1996) p.75.

minor, and a conclusion that contradicts the original *simpliciter* assertoric major. If this Bocardo is valid, then so is the original Barbara. The second part of Aristotle's proof argues that this Bocardo is valid because ordinary Bocardo is valid. The principle of inference here is that if a necessary major together with a minor implies a conclusion then that necessary major together with the minor's possibility implies the conclusion's possibility. This principle is sound. In order for the principle to apply in the present case, the following conditions would have to be met:

1. the contradictory of a *simpliciter* assertoric is a possibility-statement
2. the minor premise 'Every *c* is contingently *b*' in Barbara XQM implies a possibility-statement.

Condition 1 is met provided that a *simpliciter* assertoric 'Every *b* is *a*' has the sense 'It is necessary that every *b* is *a*' (Definition 6.9). Condition 2 is met provided that in Barbara XQM the minor premise is an 'indefinite' contingency (Definition 6.17). Aristotle's proof would then run as follows: 'If $c^\dagger \nrightarrow a^\dagger$ and $c^\dagger \rightarrow b^\dagger$ then $b^\dagger \nrightarrow a^\dagger$ (Bocardo). Therefore, in accordance with the principle that if one premise is necessary and the other possible then the conclusion is possible, if $\Box\, c^\dagger \nrightarrow a^\dagger$ and $\Diamond\, c^\dagger \rightarrow b^\dagger$ then $\Diamond\, b^\dagger \nrightarrow a^\dagger$. Therefore, by indirect reduction, if $\Box\, b^\dagger \rightarrow a^\dagger$ and $\Diamond\, c^\dagger \rightarrow b^\dagger$ then $\Diamond\, c^\dagger \rightarrow a^\dagger$.' The first premise here is implied by the *simpliciter* major '(Necessarily) every *b* is *a*' (Axiom 1.14), and the second is implied by the 'indefinite' contingency-minor 'Every *c* may or may not be *b*' (Definition 6.17). We shall therefore adopt the 'indefinite' reading of contingency-propositions for this layer.

Table 6.8 Kilwardby's 'indefinite' contingency *simpliciter* system: X/Q/M combinations

Figure	*Combinations*
1	XQM, QXM
2	XQM, QXM
3	XQM, QXM

Given the equivalence between *simpliciter* assertorics and necessity-propositions, Theorems 6.19 to 6.24 imply the following.

Theorem 6.25. The XQM moods with *simpliciter* assertorics and 'indefinite' contingencies are valid in Figure 1.

Theorem 6.26. The XQM moods with *simpliciter* assertorics and 'indefinite' contingencies are valid in Figure 2.

Theorem 6.27. The XQM moods with *simpliciter* assertorics and 'indefinite' contingencies are valid in Figure 3.
Theorem 6.28. The QXM moods with *simpliciter* assertorics and 'indefinite' contingencies are valid in Figure 1.
Theorem 6.29. The XQM moods with *simpliciter* assertorics and 'indefinite' contingencies are valid in Figure 2.
Theorem 6.30. The XQM moods with *simpliciter* assertorics and 'indefinite' contingencies are valid in Figure 3.

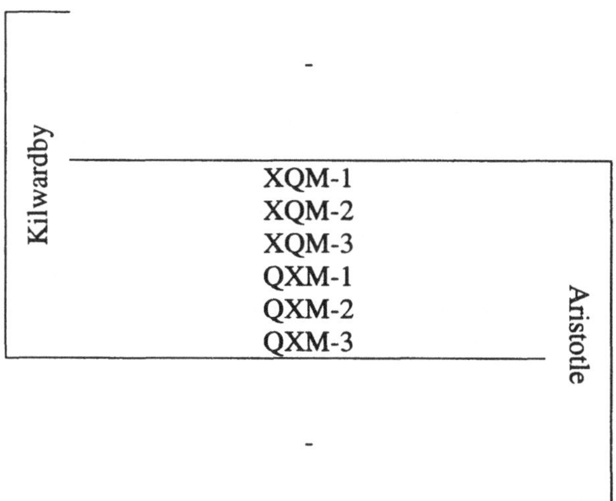

Figure 6.5 Kilwardby's 'indefinite' contingency *simpliciter* X/Q/M system and Aristotle's system

This ought to be a satisfactory result; but Kilwardby, while agreeing with Aristotle that the assertoric premise in these moods has to be *simpliciter*, wants to use the expression '*simpliciter* assertoric' more widely in this context than he does elsewhere:

> ...'*simpliciter* assertoric' is taken differently in respect of necessity-conclusions and of possibility-conclusions. A *simpliciter* assertoric that, in conjunction with a necessity-proposition, implies a necessity-conclusion, has to be necessary in fact, and it is not enough that it is a natural contingency asserted to be so, e.g. every literate of necessity has knowledge, and every man is literate (let it be so), but it is not of necessity that every man has knowledge. But in respect of a

> contingency-conclusion the assertoric may be manifested in a natural contingency....[45]

He had earlier highlighted a feature of ordinary *simpliciter* assertorics (our Axiom 1.14) upon which the validity of some syllogisms depends (see Theorem 6.2):

> The subject stands for all those things that can be under it; for the predicate possibly inheres in all those things. For, whatever can be a man is possibly an animal.[46]

This feature is shared by natural contingencies like 'Men are literate'. Here too, what possibly falls under the subject possibly falls under the predicate. So it may be that he thinks that propositions of the form '$\Box\, b^{\dagger} \rightarrow a^{\dagger}$' – not just propositions of the form '$\Box\, b \rightarrow a$' – should count as *simpliciter* in the present context.[47] But if we follow this thought, we will lose the perfect correspondence with Aristotle's X/Q/M moods.

Camestres and Baroco XQM are invalid: it is necessary that whatever is possibly seeing is possibly blind, and it is possible that nothing seeing is blind; yet it is not possible that something seeing is not seeing.

Bocardo XQM is also invalid: if $\Box\, c^{\dagger} \nrightarrow a^{\dagger}$ and $\Diamond\, c^{\dagger} \rightarrow b^{\dagger}$ then we can infer that $\Diamond\, b^{\dagger} \nrightarrow a^{\dagger}$, but not that $\Diamond\, b \nrightarrow a$.[48] It is preferable, therefore, to rest content with the ordinary definition of *simpliciter* assertorics and the perfect correspondence with Aristotle's results that it affords, rather than attempting to introduce a new understanding of *simpliciter* propositions.

Summary

Kilwardby's work is rich in insights. It is also both fragmented and elusive. Considerable interpretive stamina is required of anyone wanting to work out what Kilwardby's emerging modal theory is, as it evolves in his commentary. The commentary appears to be driven by the dual aims of constructing a viable logical theory and of showing that Aristotle was right; and these aims cannot always be reconciled with each other. Accordingly,

---

[45] Kilwardby (1516) 25rb [*ad* A15, *dub*.9].

[46] Kilwardby (1516) 24va [*ad* A15, *dub*.5].

[47] It should, however, be mentioned that this feature is shared more widely still, e.g. by propositions in which a privation is predicated of its contrary. Whatever is possibly seeing is possibly blind.

[48] Kilwardby (1516) 32va [*ad* A21, *dub*.3] consider, and rejects, a putative reduction by 'upgrading' of Bocardo XQM to Barbara LXL.

the interpretations advanced in this chapter can only be speculative. What is clear is that Kilwardby means to integrate the modal syllogistic into an Aristotelian metaphysical framework, making use of a variety of notions including that of *per se* necessity, and two different types of *simpliciter* assertoric. It is also clear that he goes beyond Averröes in dealing with modal syllogisms in all three figures, and in discussing contingency-syllogisms. His treatment is much more comprehensive in its scope, dealing with every Aristotelian combination of premises. It also contains many refinements absent from Averröes – including a differential analysis of affirmative and negative broad-sense necessity-propositions.

Kilwardby's presentation of his reflections on modal syllogistic falls into four segments – one dealing with *per se* necessities and the L/M layer; a second dealing with ampliated contingency-propositions and the Q/Q layer; a third dealing with unampliated 'natural' contingencies, *per accidens* necessities and *simpliciter* assertorics in the L/X/M, X/Q, L/Q and L/X/Q layers; and a fourth dealing with 'indefinite' contingencies in the L/Q/M and X/Q/M layers.

The first of these segments covers the L/M layer in the way mapped out by Averröes. Necessity-propositions, read as *per se*, are convertible. Kilwardby's approach validates all the Aristotelian pure necessity-syllogisms. Aristotelian methods of reduction to the first figure are available in all cases except Baroco and Bocardo.

The second segment (covering Aristotle's Q/Q layer) uses ampliated contingency-propositions in the manner proposed by Alexander, thus validating all the Aristotelian uniform contingency-syllogisms. These propositions are convertible; so Aristotelian methods of reduction to the first figure are available.

The third segment, (covering the L/X/M, X/Q, L/Q and L/X/Q layers in Aristotle) uses the broad sense of necessity-propositions along with 'natural' contingency-propositions and *simpliciter* assertorics. The semantics we have attributed to Kilwardby generate all the Aristotelian mixed L/X syllogisms, and many more besides. Since broad-sense necessity-propositions are convertible, the Aristotelian process of direct reduction is available in this layer. The X/Q and L/Q layers deliver all of Aristotle's results except for Disamis XQQ and LQQ. Being based on 'natural' contingency-propositions, which are not convertible, Kilwardby's version of this layer is not able to replicate Aristotle's direct reductions.

In the fourth layer, Kilwardby's L/Q/M and X/Q/M syllogisms exactly duplicate Aristotle's. We have attempted to provide a semantic representation of Aristotle's unique method of validating these syllogisms, using the principle that if one premise is necessary and the other possible then the conclusion is possible.

Chapter 7

Campsall

In the *Quaestiones* on the *Prior Analytics* composed by Richard Campsall (born c.1280) we encounter yet another literary form, that of the question-commentary. Of Richard's 20 questions, all but six relate to modal logic, and two of those six are also relevant, dealing as they do with the logic of tensed propositions.

Possibility

Campsall uses a range of singular terms, including ones like 'pale Socrates' that juxtapose a proper name with a denominative term. While it is necessary that Socrates is Socrates, it is contingent that pale Socrates is Socrates. Using Definition 1.6, we may conclude that 'Socrates' is a *per se* term and 'pale Socrates' is not. The crucial idea here would be that some singular terms apply to their subjects *essentially* and some do not. Within the framework of Aristotelian metaphysics names of primary substances – singular terms like 'Socrates' or 'King Kong' – apply to their subjects essentially, whereas singular terms like 'pale Socrates' or 'King Kong on the Empire State Building' do not. In our formal language this means that for each singular term 'x' there exists a corresponding starred term 'x^*', such that it may or may not be necessary that if 'x' applies to any individual then 'x^*' applies to that individual. In the former case 'x' is a name like 'Socrates'; in the latter, it is a name like 'pale Socrates'. In the former case it is necessarily true that whatever is x is x^*. In this respect singular terms standing for primary substances are formally like *per se* general terms. It is necessary that whatever is Socrates is essentially Socrates; and it is necessary that whatever is a man is essentially a man. Therefore I will extend our convention of using an upper case letter to signify a *per se* term to the case of singular terms.

Convention 7.1. 'X' abbreviates '$\Box\, x \rightarrow x^*$'.

In the case of a term that is formed by disjoining a number of singular terms – a term such as 'x or y or z' – we could adopt the convention that there always exists a corresponding starred term '$(x \text{ or } y \text{ or } z)^*$' which we stipulate is equivalent to the disjunctive term 'x^* or y^* or z^*'. Such a disjunctive term would then be a *per se* term, in our extended sense, provided that it is necessary that everything that is either x or y or z is either essentially x or essentially y or essentially z. Thus, 'Socrates or King Kong' would be a *per se* term; but 'pale Socrates or King Kong' would not since something that is either pale Socrates or King Kong (namely pale Socrates, or indeed Socrates himself) is neither essentially pale Socrates nor essentially King Kong. This construction is reflected in the following postulate, in which, we might say, 'S_a' expresses the collective individual essence of the (actual) *a*s.

Postulate 7.1. For every term 'a' there is a *per se* term 'S_a' such that $a \leftrightarrow S_a$. 'S_a' can be constructed by disjoining all the *per se* singular terms 'X', 'Y', 'Z', ...falling under 'a'.

Such terms appear to be governed by the following Axioms.

Axiom 7.1. If $S_b \mid S_a$ then $\Box\, S_b \mid S_a$.

Axiom 7.2. If $S_b \rightarrow S_a$ then $\Box\, S_b \rightarrow S_a$.

Campsall distinguishes two types of possibility-proposition. In one, the predicate stands for what can be, in the other for what is. He devotes most of his attention to the second type, and so shall we. This type of possibility-statement can be defined as follows.

Definition 7.1 (Campsall possibility). 'a' is possible for x ($x \rightarrow a^{\dagger}$) iff for no '$d$', '$e$', $x \rightarrow D$ and $\Box\, d \mid e$ and $E \leftarrow a$.

Campsall's most distinctive claim – so distinctive that we may call it Campsall's Thesis – is that an affirmative divisive possibility-proposition, expressing what I have called a Campsall possibility, implies the corresponding assertoric.[1]

Theorem 7.1 (Campsall's Thesis). If $x \rightarrow a^{\dagger}$ then $x \rightarrow a$. Suppose x is not a. Then by Postulate 7.1 there are two *per se* terms 'S_x' and 'S_a', where $S_x \leftrightarrow x$ and $S_a \leftrightarrow a$. Since x is not a, $S_x \mid S_a$. Hence, by Axiom 7.1, $\Box\, S_x \mid S_a$. So we have $x \rightarrow S_x$ and $\Box\, S_x \mid S_a$ and $S_a \leftarrow a$, which by Definition 7.1 implies that x is not possibly a. This thesis, whereby possibility-propositions are assimilated to assertorics, recalls Avicenna's assimilation of assertorics to possibility-propositions. But, as we shall see, Avicenna's modal logic is quite different from Campsall's.

Similar considerations apply to future-tense statements. The governing principle (which Campsall makes explicit) is that:

> ...if it is true to say of something that it will be one of those things that are now contained under 'a', then it is now contained under 'a'.[2]

---

[1] Campsall (1968) 19.21.

[2] Campsall (1968) 4.44-4.45.

The principles relied on in relation to statements of possibility or futurity are clarified by his views about a topic that Abélard had discussed, namely the possibility of being a bishop:

> It is to be said that 'It is possible for you to be a bishop' in the divided sense is to be further distinguished, in that the predicate can be taken for things that are or else for things that can be. If the first way, the inference is good from that to the assertoric 'You are a bishop', because the antecedent in that sense signifies that you can be some bishop existing now, and this cannot be true unless you are now a bishop.[3]

This discussion stands in stark contrast with Abélard's treatment of the question. According to Abélard, a sufficient condition for the existence of this possibility in relation to a given person, is that it be actualized for *some* person; whereas according to Campsall, a necessary and sufficient condition for the existence of the possibility is that it be actualized for the particular person under discussion. We can elaborate on this idea as follows. If George is the only bishop now in existence and if Socrates is not George then (speaking of actual bishops) it is not possible for Socrates to be a bishop – the reason being that it is not possible for Socrates to be George. Impossibility therefore, for Campsall, does not rest, as it did for Abélard, on an incompatibility between a predicate and a general nature. If it did then impossibilities would occur in whole species together. What is apparent from Campsall's discussion is that what is impossible for one member of a species may be possible, indeed may be actual, for other members of the same species.

By contrast with possibility-propositions *de esse*, possibility-propositions *de non esse* do not, according to Campsall, imply the corresponding assertorics. If George can be a non-bishop, it does not follow that he is not a bishop. Campsall himself states the result in this form, when he says that it is not valid to infer from a negative possibility-proposition to its corresponding negative assertoric:

> ...this rule should not be understood to apply to negative possibility-propositions, because they never imply the corresponding assertorics, unless by virtue of the meaning of their terms.[4]

Lagerlund points out that if Campsall had not given this differential treatment of possibilities *de esse* and *de non esse*, he would have had to

[3] Campsall (1968) 19.23.
[4] Campsall (1968) 19.21.

face the consequence that contingency-propositions – which imply two possibilities, one *de esse* and one *de non esse* – are self-contradictory.[5] We add that Campsall seems to be denying that we can read the predicates of possibility-statements *de non esse* as standing for what is. (If we could, then 'George can be a non-bishop' would appear to be false if George is a bishop, since in that case there is no non-bishop that George can be.)

Campsall holds that universal negative apodeictics have two senses – compound and divided. In the compound sense, we have a singular proposition predicating necessity of an 'inherence'. In this sense the proposition is convertible, because in general if the antecedent is necessary so is the consequent. In the divided sense we have a universal proposition signifying that 'anything contained under the predicate is necessarily removed from anything contained under the subject'. In this sense, he says, it is also convertible, provided that both predicate and subject stand for their respective *supposita*, i.e. those things that now fall under them, not what will do so or has done so.[6] We are now in a position to define modal propositions *de esse* as Campsall understands them.

Definition 7.2 (Ma). 'Every b is possibly a': $b \rightarrow a^{t}$. 'Every b is possibly a' in this sense is equivalent to 'Every b is a' (Axiom 7.1).

Definition 7.3 (Le). 'No b is possibly a': for some 'd', 'e', $b \rightarrow D$ and $\Box\, d \mid e$ and $E \leftarrow a$. 'No b is possibly a' in this sense is equivalent to 'No b is a' (Postulate 7.1, Axiom 7.1).

Definition 7.4 (Mi). 'Some b is possibly a': $b \smile a^{t}$. 'Some b is possibly a' in this sense is equivalent to 'Some b is a' (Axiom 7.1).

Definition 7.5 (Lo). 'Some b is not possibly a': for some 'd', 'e', 'f', $b \leftarrow D$ and $D \rightarrow E$ and $\Box\, e \mid f$ and $F \leftarrow a$. 'Some b is not possibly a' is equivalent to 'Some b is not a' (Postulate 7.1, Axiom 7.1). It emerges from Definitions 7.2 to 7.5 that the *de rebus* modal propositions in Abélard's *de esse* square of opposition (Ma, Le, Mi, Lo) are, on Campsall's reading, all equivalent to the corresponding assertorics.

Necessity

Campsall does not follow Kilwardby's usage according to which an assertoric that is in fact necessary counts as *simpliciter* (Definitions 6.9 to 6.12). Kilwardby had explicitly included 'Whatever is walking is moving' in this class. Instead, Campsall divides assertoric propositions into those

---

[5] Lagerlund (2000) p.72.

[6] Campsall (1968) 5.38-5.40.

that are assertoric *simpliciter* (these are unconditionally necessary), those that are *ut nunc* in the first way (these – propositions like 'Some man is living', 'Socrates is a man', 'Socrates is living' – are contingent but their terms are mutually invariant, conditionally on the existence of the subject), and those that are *ut nunc* in the second way (these – propositions like 'A white man is a man' – are contingent, with variable terms, even if the terms are inseparable).[7] He says that the latter type of proposition is excluded from Aristotle's mixed modal syllogisms.[8]

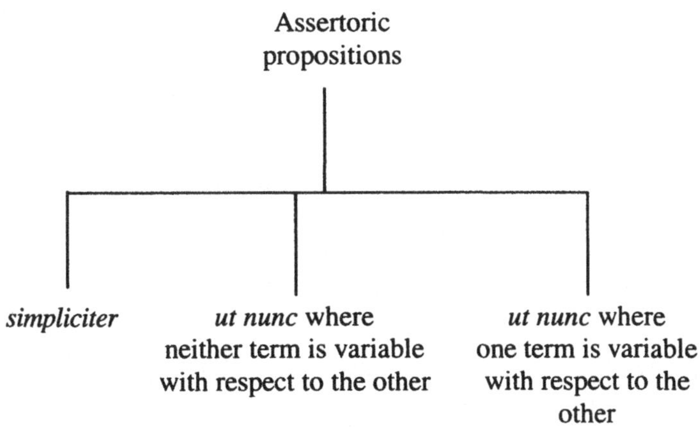

Figure 7.1 Campsall's classification of assertorics

Campsall appears to be distinguishing three aspects of necessity – unconditionality, invariance, and inseparability (of predicate and subject). *Simpliciter* assertorics possess all three of these characteristics. Lagerlund takes the unconditionality of these propositions to reside in the fact that they are 'about always existing individuals'.[9] Other types of statement may have non-eternal subjects and are conditional on the existence of those subjects. *Ut nunc* assertorics of the first type, though merely conditional, must express invariability and inseparability. *Ut nunc* assertorics of the second type may have terms that are mutually variable. Their predicates may be inseparable from their subjects, but (if the classification is to be complete) these propositions may lack even this vestige of necessity.

The notion of inseparability seems to be captured in the notion of *de dicto* inclusion. Invariability is a more elusive notion. Perhaps we can say

[7] Campsall (1968) 12.31.
[8] Campsall (1968) 12.32.
[9] Lagerlund (2000) p.86.

that a term '*b*' is invariable with respect to a term '*a*' provided that every *b* either must or cannot be *a*, and every *a* either must or cannot be *b*. Invariability occurs with species of a genus: 'man' is invariable with respect to 'animal' because being an animal is necessary for every man and being a man is either necessary or impossible for every animal. Here, both terms are *per se*. (Campsall gives some examples of mutually variable terms, among them the pair 'white man', 'animal'.) But it also seems to be the case that a term like 'white', that is necessarily implied by a *per se* term like 'snow', is invariant with respect to the latter term. All snow must be white, and everything that is white either must be snow or cannot be snow. If that is right, affirmative invariant propositions are those whose subject is *per se* and whose predicate is necessarily implied by that subject.

Given that even second-class *ut nunc* propositions are in a sense necessary, we have to ask what, in the context of modal syllogistic, Campsall means by a universal affirmative necessity-proposition. For him universal affirmative divided necessity-propositions possess the feature of invariability; however, these propositions need not be unconditional, as the following passage (speaking of *particular* propositions) shows.

> An affirmative necessity-proposition in the divided sense is true only if something contained under the predicate term is invariant with respect to something under the subject term when it exists....[10]

Thus, Campsall's truth-conditions for affirmative necessity-propositions appear to be the same as for the class of propositions that are either *simpliciter* or *ut nunc* in the first way. The one condition required in both cases is invariance of the terms relative to each other, i.e. either necessary inclusion or necessary exclusion of a predicate relative to a *per se* subject. This definition agrees with most of Campsall's examples. For instance, he holds as true the propositions 'Socrates of necessity is living', or 'mortal', or 'exists',[11] where all of these predicates are necessitated by the subject 'Socrates' provided of course that we understand that subject as standing for a human being. (On the other hand, he says that Socrates is of necessity blind, given that he is blind;[12] but this can only be a relative necessity.) Given this, I conjecture the following definitions of Campsall's *de non esse* modal propositions.

Definition 7.6 (La). 'Every *b* is necessarily *a*': $\Box B \rightarrow a$. If every *b* is necessarily *a* in this sense then every *b* is *a*.

---

[10] Campsall (1968) 6.25.
[11] Campsall (1968) 5.44.
[12] Campsall (1968) 5.46.

Definition 7.7 (Me). 'No b is necessarily a': for every *per se* term 'd', if $\Box D \rightarrow b$ then $\sim\Box D \rightarrow a$. We can abbreviate this to say that no *per se suppositum* of 'b' is a *per se suppositum* of 'a'. If no b is a then no b is necessarily a in this sense.
Definition 7.8 (Li). 'Some b is necessarily a': for some *per se* term 'd', $\Box D \rightarrow b$ and $\Box D \rightarrow a$. In other words, some *per se suppositum* of 'b' is a *per se suppositum* of 'a'. If some b is necessarily a in this sense then some b is a.
Definition 7.9 (Mo). 'Some b is not necessarily a': either $\sim\Box b \rightarrow b^*$ or for some *per se* term 'd', $\Box D \rightarrow b$ and $\sim\Box D \rightarrow a$. If some b is not a then some b is not necessarily a in this sense.

Necessity-syllogisms

L/M combinations

Campsall does not discuss them explicitly, but the Aristotelian pure necessity-syllogisms are all valid on his reading. Moreover, their validity is derivative upon that of non-modal syllogisms, given Rule 1.3 and modal subordination (Definitions 7.3, 7.5, 7.6 and 7.8).
[Theorem 7.2. The LLL moods are valid in Figure 1.] Barbara: if $\Box b \rightarrow a$ and $\Box c \rightarrow b$ then $\Box c \rightarrow a$ (applying Rule 1.3 to Axiom 1.3), so if $\Box B \rightarrow a$ and $\Box C \rightarrow b$ then $\Box C \rightarrow a$. Darii can be demonstrated in a similar way. Celarent and Ferio LLL are equivalent, by Definitions 7.3 and 7.5, with Celarent and Ferio XLX, whose validity follows from that of assertoric Celarent and Ferio by virtue of modal subordination (Definitions 7.6 and 7.8).

We have two corollaries by indirect reduction.
[Theorem 7.2a. The LMM moods are valid in Figure 2.]
[Theorem 7.2b. The MLM moods are valid in Figure 3.]
[Theorem 7.3. The LLL moods are valid in Figure 2.] These are equivalent, by Definitions 7.3 or Theorem 7.5, with XLX or LXX second-figure moods, whose validity follows by modal subordination from that of their assertoric counterparts.

Again there are two corollaries.
[Theorem 7.3a. The LMM moods are valid in Figure 1.]
[Theorem 7.3b. The LMM moods are valid in Figure 3.]
[Theorem 7.4. The LLL moods are valid in Figure 3.] Proofs as for Theorem 7.3.
[Theorem 7.4a. The MLM moods are valid in Figure 1.]

[Theorem 7.4b. The MLM moods are valid in Figure 2.]

Table 7.1 Campsall's system: L/M combinations

Figure	*Combinations*
1	LLL, LMM, MLM, Celarent Ferio LML, Barbara Darii MMM
2	LLL, LMM, MLM, Cesare Festino LML, Camestres Baroco MLL
3	LLL, LMM, MLM, Ferison Bocardo LML, Datisi Disamis MMM

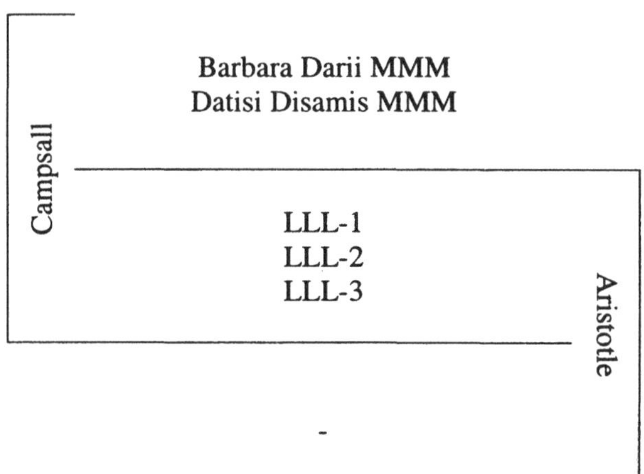

Figure 7.2 Campsall's L/M system and Aristotle's system

In addition, given our truth-conditions for possibility-propositions, and given Campsall's Thesis, we have the following theorems.
[Theorem 7.5. Barbara and Darii MMM are valid.] Barbara and Darii MMM are equivalent, by Definitions 7.2 and 7.4, with non-modal Barbara and Darii.
[Theorem 7.5a. Camestres and Baroco MLL are valid.]
[Theorem 7.5b. Ferison and Bocardo LML are valid.]
[Theorem 7.6. Datisi and Disamis MMM are valid.] Datisi and Disamis MMM are equivalent, by Definitions 7.2 and 7.4, with non-modal Datisi and Disamis.
[Theorem 7.6a. Celarent and Ferio LML are valid.]
[Theorem 7.6b. Cesare and Festino LML are valid.]

The picture that emerges is shown in Table 7.1 and Figure 7.2. Campsall's set of syllogisms is more comprehensive than Aristotle's, but not as comprehensive as Avicenna's (Table 4.1).

L/X/M combinations

Here again the validity of the valid syllogisms is derivative upon that of non-modal syllogisms. With ordinary assertorics, there are no valid affirmative mixed necessity-assertoric syllogisms in the first figure. If $\Box B \rightarrow a$ and $c \rightarrow b$, all we may conclude is that $c \rightarrow a$. No necessity-conclusion follows. Campsall however does accept some L/X/M combinations in Barbara. Let us start with the affirmative moods:

> There is a syllogism in the first figure from an assertoric major and a possibility-minor to a possibility-conclusion in affirmative moods, and it is impossible for the inference to fail there, because it is impossible to find terms such that the conclusion is false when the premises are true in the divided sense. Nor is it valid to say that this syllogism is not valid because Aristotle only asserts the combination of an assertoric and a two-way contingency-proposition – because, just as the combination of an assertoric and a contingency-proposition can be proved by arguing from the opposite of the conclusion and making the minor an assertoric, in the same way when the minor is a possibility-proposition the combination can be proved by arguing from the opposite of the conclusion and making the possibility-minor an assertoric.[13]

As noted in Chapter 1,[14] the procedure whereby Aristotle claims to validate Barbara XQM is not self-evidently sound. It is not clear what warrants the 'upgrading' of a contingency-proposition to an assertoric in that procedure. We proposed an interpretation of this procedure in relation to Kilwardby's logic.[15] In Campsall's logic, the problem disappears because for him there is no semantic difference between an affirmative possibility-proposition and the corresponding assertoric. Thus, it is always legitimate to replace an affirmative possibility-proposition by the corresponding assertoric. There is an irony here. Campsall is saying that Barbara XMM can be proved in the same way that Aristotle proves Barbara XQM, but he does not mention the fact that Aristotle requires the major premise in that syllogism to be *simpliciter*. In any event, since first-figure XXX moods are valid and affirmative assertorics are equivalent to the corresponding possibility-propositions, we have the following theorem.

---

[13] Campsall (1968) 14.16-17.

[14] See pp.15-16 above.

[15] See pp.111-112 above.

Theorem 7.7. The XMM moods are valid in Figure 1. Barbara and Darii are equivalent, by Definitions 7.2 and 7.4, with their non-modal counterparts. Celarent and Ferio are equivalent by Definition 7.2 to XXM moods, which reduce to assertoric moods by courtesy of modal subordination (Definitions 7.7 and 7.9).

Campsall states that the validity of Baroco XLL follows from this; and to this we add the other second-figure moods.

Theorem 7.7a. The XLL moods are valid in Figure 2.[16]

Also equivalent to first-figure XMM moods are the following.

[**Theorem 7.7b.** The LMX moods are valid in Figure 3.] Among the syllogisms asserted to be valid by Theorem 7.7 and its corollaries, Campsall states that Baroco XLL is syllogistic. This is correct, and applies to all the syllogisms in this group because all are equivalent to assertoric syllogisms.

Campsall states the validity of Barbara LMX, and it is easy to see that the other first-figure LMX moods are valid (all being equivalent, by Definitions 7.2 and 7.4, to LXX moods), so we have:

Theorem 7.8. The LMX moods are valid in Figure 1.[17]

Campsall names some of the equivalent moods in Figure 2, but we may draw the general inference:

Theorem 7.8a. The LXL moods are valid in Figure 2.[18] Among the syllogisms in this group, Campsall states that Baroco LXL, while valid, is not syllogistic. This is correct, because its validity reduces to that of assertoric Baroco only after the replacement of '$\Box B \rightarrow a$' by '$b \rightarrow a$'. The same applies to all the syllogisms in this group.

We add the equivalent third-figure moods:

[**Theorem 7.8b.** The XMM are valid in Figure 3.]

Along with Barbara XMM and LMX, Campsall accepts Barbara MXM and MLX. To these we may add the corresponding versions of Darii.

Theorem 7.9. Barbara and Darii MXM are valid.[19] These are equivalent, by Definitions 7.2 and 7.4, with the corresponding assertoric moods, and so are syllogistically valid.

By contrast, the negative moods are invalid.

Theorem 7.10. Barbara and Darii MLX are valid.[20] These are equivalent to the analogous XLX moods, which in turn are strengthened versions of non-modal syllogisms. They are not syllogistically valid (and Campsall notes

---

[16] Campsall (1968) 14.16.
[17] Campsall (1968) 14.18.
[18] Campsall (1968) 14.18.
[19] Campsall (1968) 14.19.
[20] Campsall (1968) 14.19.

this),[21] because a syllogistic chain is generated only after we replace the minor premise '$\Box\, C \rightarrow b$' by '$c \rightarrow b$', forming the chain '$c \rightarrow b \rightarrow a$'.

Celarent and Ferio MLX are invalid.

Campsall draws the inference that these moods generate equivalents in the third figure:

Theorem 7.9a. Bocardo and Ferison LXL are valid.[22]

Theorem 7.10a. Bocardo and Ferison XLL are valid.[23]

To them we may add pairs of equivalent syllogisms in Figure 2.

[**Theorem 7.9b.** Baroco and Camestres MLX are valid.]

[**Theorem 7.10b.** Baroco and Camestres MXM are valid.]

Let us turn to the negative moods in Figure 1. He does not say whether Celarent LXL is valid with a minor premise that is *ut nunc* in the second way, but we can easily see that, given the equivalence of universal negative necessity-propositions with their corresponding assertorics, the validity of Celarent LXL is equivalent to that of Celarent XXX.

[**Theorem 7.11.** Celarent and Ferio LXL are valid.] These moods are syllogistic, being equivalent by Definition 7.3 and 7.5 to non-modal syllogisms.

Barbara and Darii LXL are invalid. Campsall gives the counter-example 'Every man of necessity is an animal, a pale man is a man, so a pale man of necessity is an animal'.[24] The example echoes one of Kilwardby's.[25]

Celarent and Ferio LXL are equivalent, by indirect reduction, to moods in the other figures, so that we have the following theorems.

[**Theorem 7.11a.** Festino and Cesare LMX are valid.]

[**Theorem 7.11b.** Disamis and Datisi MXM are valid.]

In the case of the XLL moods, Campsall states that the mixture of a negative assertoric major with an apodeictic minor implies an apodeictic conclusion, whether the assertoric premise is *simpliciter* or *ut nunc* in either of the two ways he distinguishes.[26] The reason he gives is that a negative assertoric implies the corresponding negative necessity-proposition. Thus he is saying that because Celarent XLX is valid, so is Celarent XLL. Celarent XLX, however, is not syllogistically valid, since its validity depends on a strengthened minor.

Theorem 7.12. Celarent and Ferio XLL are valid. By Definition 7.3 these moods are equivalent to the corresponding LLL moods, which are valid

---

[21] Campsall (1968) 14.19.
[22] Campsall (1968) 14.19.
[23] *Ibid.*
[24] Campsall (1968) 12.32.
[25] Kilwardby (1516) 16vb [*ad* A9, *dub.*6].
[26] Campsall (1968) 13.15.

(Theorem 7.2). Barbara and Darii XLL can be shown invalid by Aristotle's counter-example.

The equivalent second- and third-figure moods are identified in the following two theorems.

[**Theorem 7.12a.** Festino and Cesare XMM are valid.]

[**Theorem 7.12b.** Disamis and Datisi MLX are valid.]

Table 7.2 Campsall's system: L/X/M combinations

Figure	*Combinations*
1	LMX, XMM, Celarent Ferio LXL, Celarent Ferio XLL, Barbara Darii MLX, Barbara Darii MXM,
2	LXL, XLL, Cesare Festino LMX, Cesare Festino XMM, Camestres Baroco MLX, Camestres Baroco MXM
3	LMX, XMM, Ferison Bocardo LXL, Ferison Bocardo XLL, Datisi Disamis MLX, Datisi Disamis MXM

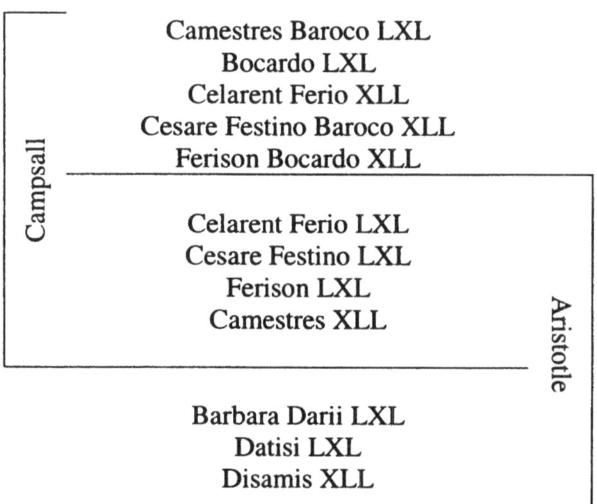

Figure 7.3 Campsall's L/X/M system and Aristotle's system

Figure 7.3 compares these results with those of Aristotle, as displayed in Table 1.4. As can be seen, there is a common core of syllogisms, along with a set of syllogisms peculiar to Campsall and one peculiar to Aristotle. The moods peculiar to Campsall include Camestres and Baroco and Bocardo LXL, along with Celarent, Ferio, Baroco, Ferison and Bocardo

XLL; those peculiar to Aristotle include Barbara and Darii LXL, Datisi LXL and Disamis XLL.

The results are different if assertorics are *simpliciter* or *ut nunc* in the first way. Having carefully distinguished between *simpliciter* assertorics and first-class *ut nunc* propositions, Campsall effectively merges these two types in his modal syllogistic. As noted earlier, a universal proposition that is either *simpliciter* or *ut nunc* in the first way has the same truth-conditions, according to Campsall, as a proposition expressing a necessary inclusion or exclusion relative to a *per se* subject. Accordingly, we adopt the following definitions.

Definition 7.10 (Xa *simpliciter* or first-class *ut nunc*). 'Every b is a': $\Box B \rightarrow a$.

Definition 7.11 (Xe *simpliciter* or first-class *ut nunc*). 'No b is a': For some 'd', 'e', $b \rightarrow d$ and $\Box D \mid E$ and $a \rightarrow e$.

Definition 7.12 (Xi *simpliciter* or first-class *ut nunc*). 'Some b is a': For some 'd', $\Box B \leftarrow d$ and $\Box A \leftarrow d$.

Definition 7.13 (Xo *simpliciter* or first-class *ut nunc*). 'Some b is not a': For some 'd', 'e', 'f', $b \rightarrow d \rightarrow e$ and $\Box E \mid F$ and $a \rightarrow f$.

It is clear from these definitions and Definitions 7.6 to 7.9 that *de non esse* modals are equivalent to propositions that are *simpliciter* or *ut nunc* in the first way, La and Li being equivalent to Xa and Xi, Me and Mo to Xe and Xo. On the other hand, as we saw earlier, *de esse* modals are equivalent to ordinary assertorics, Ma and Mi being equivalent to Xa and Xi, Le and Lo to Xe and Xi. Therefore, all modal propositions are equivalent, in Campsall's scheme, to non-modal propositions, either to ordinary assertorics or to propositions that are *simpliciter* or *ut nunc* in the first way.

Theorem 7.13. The first-figure LXL moods are valid if the assertoric premise is *simpliciter* or *ut nunc* in the first way.[27] The affirmative moods are equivalent to LLL, the negative to XLX moods (with ordinary assertorics).

By indirect reduction, these moods have equivalents in Figures 2 and 3:

[**Theorem 7.13a.** The third-figure MXM moods are valid on the same reading.]

Theorem 7.14. The first-figure XLL moods are valid on the same reading.[28] The affirmative moods are equivalent to LLL moods. Campsall thinks these moods, though valid, are not syllogistic.[29] The negative moods are equivalent to XLX moods (with ordinary assertorics).

---

[27] Campsall (1968) 13.32.

[28] Campsall (1968) 13.35.

[29] Campsall (1968) 12.34-35; 13.14.

By indirect reduction we have:
[Theorem 7.14a. The second-figure XMM moods are valid on the same reading.]

Table 7.3 Campsall's *simpliciter* (or first-class *ut nunc*) system: L/X/M combinations

Figure	*Combinations*
1	LXL, XLL, XMM, MXM
2	LXL, XLL, XMM, MXM
3	LXL, XLL, XMM, MXM

[Theorem 7.15. The second-figure LXL moods are valid on the same reading.] Those with negative major are equivalent to XLX moods. Those with affirmative major are equivalent to LXX moods (with ordinary assertorics).
[Theorem 7.15a. The third-figure XMM moods are valid on the same reading.]
[Theorem 7.16. The second-figure XLL moods are valid on the same reading.] Those with affirmative major are equivalent to LXX moods. Those with negative major are equivalent to XLX moods (with ordinary assertorics).
[Theorem 7.16a. The first-figure XMM moods are valid on the same reading.]
[Theorem 7.17. The LXL moods in the third figure are valid on the same reading.] The affirmative moods are equivalent to LLL, the negative to XLX moods (with ordinary assertorics).
[Theorem 7.17a. The MXM moods in the first figure are valid on the same reading.]
[Theorem 7.18. The third-figure XLL moods are valid.] Those with affirmative major are equivalent to LLL moods. Those with negative major are equivalent to XLX moods (with ordinary assertorics).
[Theorem 7.18a. The second-figure MXM moods are valid on the same reading.]

These results are summarized in Table 7.4. Comparison with Table 6.1 shows that this is identical with Kilwardby's maximal L/X/M system.

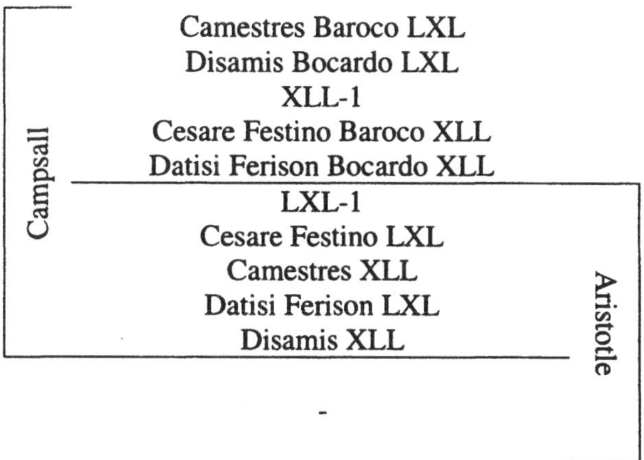

Figure 7.4 Campsall's L/X/M *simpliciter* (or first-class *ut nunc*) system and Aristotle's system

Contingency-syllogisms

Campsall explains '*b* may or may not be *a*' in the divided sense as stating that 'something contained under the predicate is in something contained under the subject'.

> ...and so for its truth in the divisive sense it is required that this *b* may or may not be this *a* or that *a*, and so on for the others. Accordingly, propositions like 'A man may or may not be an animal' are to be denied in the divisive sense.[30]

He explains that '*b* may or may not be an *a*' means not '*b* can be an *a* and can not be an *a*' but '*b* can be an *a* and can not be that *a*'.[31] Similarly, 'It is contingent that *b* is not an *a*' means 'it is possible that *b* is an *a* and it is possible that that *b* is not that *a*'.[32]

Q/Q combinations

Campsall accepts the pure contingency-syllogisms of the first and third figures in the divided sense.[33] Like Kilwardby, he makes use of both an

[30] Campsall (1968) 7.34.
[31] Campsall (1968) 7.35.
[32] Campsall (1968) 16.23.
[33] Campsall (1968) 17.47.

ampliated and an unampliated reading of contingency-propositions, the ampliated reading being as in definitions 6.17 and 6.18. In relation to what he takes to be the Aristotelian view that contingency-propositions must be ampliated in pure contingency-syllogisms, he introduces a refinement:

> Concerning uniform contingency-syllogisms, it is to be noted that Aristotle's statement that contingency-propositions occur in uniform syllogisms only if taken in the sense 'what may or may not be'... is to be understood as being about the major in the first figure, because the major's subject has to be taken for those things that may or may not be.... And so, in any figure, in order for a uniform contingency-syllogism to be valid, the middle term has to be taken for those things that may or may not be.[34]

In other words, the minimum requirement is that the middle term be ampliated. Consistent with this, in the first figure the minor term might be either ampliated or unampliated.

Theorem 7.19. The QQQ moods are valid in Figure 1.[35] With ampliated minor, Barbara (Celarent) is valid: if $b^\ddagger \rightarrow a^\ddagger$ and $c^\ddagger \rightarrow b^\ddagger$ then we have the chain '$c^\ddagger \rightarrow b^\ddagger \rightarrow a^\ddagger$', simplifying to '$c^\ddagger \rightarrow a^\ddagger$'. Similarly Darii (Ferio) is valid: if $b^\ddagger \rightarrow a^\ddagger$ and $c^\ddagger \smile b^\ddagger$ then we have the chain '$c^\ddagger \smile b^\ddagger \rightarrow a^\ddagger$', simplifying to '$c^\ddagger \smile a^\ddagger$'. With unampliated minor Barbara is still valid: if $b^\ddagger \rightarrow a^\ddagger$ and $c \rightarrow b^\ddagger$ then we have the chain '$c \rightarrow b^\ddagger \rightarrow a^\ddagger$', simplifying to '$c \rightarrow a^\ddagger$'. Likewise Darii (Ferio) is valid: if $b^\ddagger \rightarrow a^\ddagger$ and $c \smile b^\ddagger$ then we have the chain '$c \smile b^\ddagger \rightarrow a^\ddagger$', simplifying to '$c \smile a^\ddagger$'.

Table 7.4 Campsall's system: Q/Q combinations

Figure	*Combinations*
1	QQQ
2	-
3	QQQ

Theorem 7.20. The QQQ moods are valid in Figure 3.[36] Datisi (Ferison) is valid with ampliated middle: if $c^\ddagger \rightarrow a^\ddagger$ and $c^\ddagger \smile b^\ddagger$ then we have the chain '$b^\ddagger \smile c^\ddagger \rightarrow a^\ddagger$', simplifying to '$b^\ddagger \smile a^\ddagger$'. Disamis (Bocardo): if $c^\ddagger \smile a^\ddagger$ and $c^\ddagger \rightarrow b^\ddagger$ then we have the chain '$b^\ddagger \leftarrow c^\ddagger \smile a^\ddagger$', simplifying to '$b^\ddagger \smile a^\ddagger$'.

---

[34] Campsall (1968) 17.70.

[35] Campsall (1968) 17.48.

[36] *Ibid.*

X/Q combinations

Campsall accepts all QXQ moods in the first and third figure, provided that the assertoric minor is *simpliciter* or *ut nunc* in the first way. These results are consistent with the following definitions.

Definition 7.14 (Qa, Qe – unampliated). 'Every *b* may or may not be *a*': $\underline{b} \rightarrow a$ and no *per se suppositum* of '*b*' is a *per se suppositum* of '*a*'. This is equivalent to the conjunction of an Ma- and an Me-proposition. Because the Qa-proposition implies the corresponding Ma-proposition, it implies the corresponding universal affirmative assertoric (Definition 7.2). However, the fact that it implies the corresponding Me-proposition does not entail that it implies the corresponding universal negative assertoric.

Definition 7.15 (Qi, Qo – unampliated). 'Some *b* may or may not be *a*': for some *per se* term '*d*', $D \rightarrow a$ and $\Box\, D \rightarrow b$ and $\sim\Box\, D \rightarrow a$. Particular contingency-propositions, so defined, imply the corresponding particular possibility-propositions, both affirmative and negative. (The affirmative states that for some '*d*', $b \leftarrow d \rightarrow a$, and the negative states that either $\sim\Box\, b \rightarrow b^*$ or for some *per se* term '*d*', $\Box\, D \rightarrow b$ and $\sim\Box\, D \rightarrow a$.)

Theorem 7.21. The QXQ moods are valid in Figure 1.[37] Barbara (Celarent): if $b \rightarrow a$, and no *per se suppositum* of '*b*' is a *per se suppositum* of '*a*', and $\Box\, C \rightarrow b$, then on the one hand $c \rightarrow b \rightarrow a$ so that $c \rightarrow a$; and on the other hand, since $\Box\, C \rightarrow b$, every *per se suppositum* of '*c*' is a *per se suppositum* of '*b*', and consequently no *per se suppositum* of '*c*' is a *per se suppositum* of '*a*'. In other words, this mood reduces to Barbara MLM (XLX) and Celarent MLM. Similarly for Darii (Ferio).

[**Theorem 7.22.** The QXQ moods are valid in Figure 3.] Datisi (Ferison) reduces to Datisi MLM (XLX) and Ferison MLM. Disamis (Bocardo) reduces to Disamis MLM (XLX) and Bocardo MLM.

Campsall appears to endorse Barbara and Darii XQQ with *simpliciter* or first-class *ut nunc* major. He cites the syllogism 'Every man is Socrates, it may or may not be that Socrates is a man, so it may or may not be that Socrates is Socrates', saying that if the predicate of the minor is taken for those things that are, then the premises are mutually inconsistent.[38] What he says is true of this argument, since (as he himself says elsewhere) the minor premise is impossible, if its terms stand for what is.[39] However, that comment does not apply universally to the minor premise in these moods. A proposition like 'Every man may or may not run' may (as Campsall

---

[37] Campsall (1968) 18.37-38.

[38] Campsall (1968) 18.54.

[39] Campsall (1968) 8.24.

recognizes) be true.[40] But, a Barbara XQQ with this proposition as minor may lead from truths to a falsehood: given that everything running is a man, and that every man may or may not run, it does not follow that every man may or may not be a man.

Table 7.5 Campsall's *simpliciter* (or first-class *ut nunc*) system: X/Q combinations

Figure	*Combinations*
1	QXQ
2	-
3	QXQ

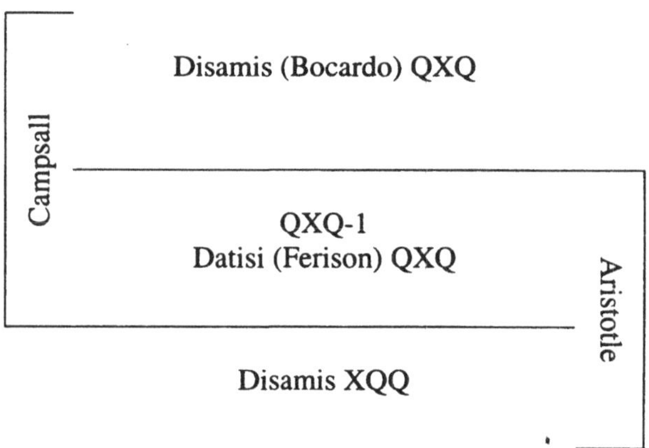

Figure 7.5 Campsall's X/Q *simpliciter* (or first-class *ut nunc*) system and Aristotle's system

The combination of a contingency major and a second-class *ut nunc* ordinary assertoric minor never yields a contingency-conclusion.[41] Campsall gives several counter-examples. The first of these argues 'Every man may or may not run, and everything running is a man, so everything running may or may not be running'.[42] In another counter-example, the premises are 'Every man may or may not be this white man or that white man,' and 'This white man is a man', and the conclusion is 'So this white

[40] Campsall (1968) 18.38.
[41] Campsall (1968) 18.38.
[42] Campsall (1968) 18.01, 18.46.

man may or may not be this white man or that white man'. Campsall states that the premises are true and the conclusion impossible. Campsall's third counter-example is 'Socrates may or may not be pale Socrates, pale Socrates is Socrates, so pale Socrates may or may not be pale Socrates'.[43] These counter-examples constitute a difficulty for our interpretation, since they are not consistent with Definitions 7.14 and 7.15. The conclusions of the examples come out as true, not false, on our proposed semantics. For instance, it is true that everything running may or may not be running, because everything running is running, and no *per se suppositum* of 'running' is a *per se suppositum* of 'running' (since there is no *per se suppositum* of 'running').

L/Q combinations

Table 7.6 Campsall's system: L/Q combinations

Figure	*Combinations*
1	QLQ
2	-
3	QLQ

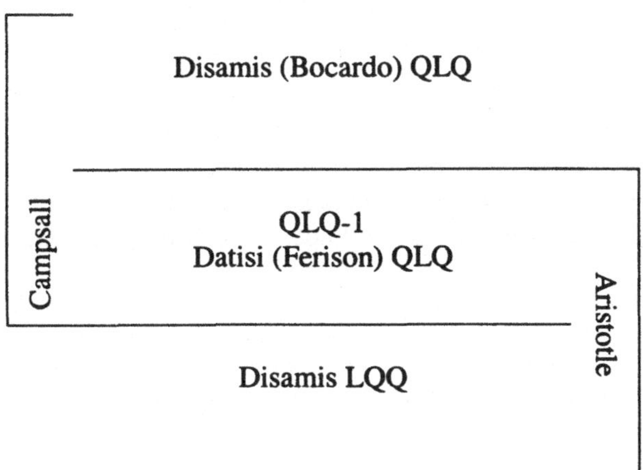

Figure 7.6 Campsall's L/Q system and Aristotle's system

Given the equivalence of necessity-propositions with assertorics that are either *simpliciter* or first-class *ut nunc*, the moods in this layer are

[43] Campsall (1968) 20.22.

equivalent to the X/Q combinations. Though not noticed expressly by Campsall, they can easily be shown to be valid in accordance with his general approach.
[**Theorem 7.23.** The QLQ moods are valid in Figure 1.]
[**Theorem 7.24.** The QLQ moods are valid in Figure 3.]

L/X/Q combinations

The syllogisms in this layer derive their validity from that of syllogisms in the L/X/M layer.
Theorem 7.25. The LQX moods are valid in Figure 1.[44]
Theorem 7.26. The LQX moods are valid in Figure 2.[45]
Theorem 7.27. The LQX moods are valid in Figure 3.[46]

Campsall's reason for accepting the LQX moods is that the contingency-minor implies the corresponding assertoric, so that wherever we have a valid LXX syllogism we have the corresponding LQX syllogism.[47] This implies the acceptance of all LQX syllogisms wherever there is a corresponding valid assertoric mood. Similar reasoning shows that all QLX syllogisms in the standard non-modal moods are valid provided that the major's predicate stands for things that are.
[**Theorem 7.28.** The QLX moods are valid in Figure 1.]
[**Theorem 7.29.** The QLX moods are valid in Figure 2.]
[**Theorem 7.30.** The QLX moods are valid in Figure 3.]

Campsall's system is a maximal one, including all LQX and QLX moods. It is broader than Aristotle's system (Table 1.8), as Figure 7.7 demonstrates. One of the salient differences is that Aristotle accepts LQX in Figure 1 only in negative moods, whereas Campsall accepts the affirmative moods as well. Campsall notes this difference, saying:

> In as much as Aristotle concedes one and denies the other, this is because one can be perfected by converting the major, and the other cannot.... And so he does not deny that this mixture is necessary, but in as much as it cannot equally evidently be proved as the negative mixture; and so he denies that it is evident, not that it is necessary.[48]

---

[44] *Ibid.*
[45] *Ibid.*
[46] Campsall (1968) 20.14-16.
[47] Campsall (1968) 20.15-16.
[48] Campsall (1968) 20.17.

Table 7.7 Campsall's system: L/X/Q combinations

Figure	*Combinations*
1	LQX, QLX
2	LQX, QLX
3	LQX, QLX

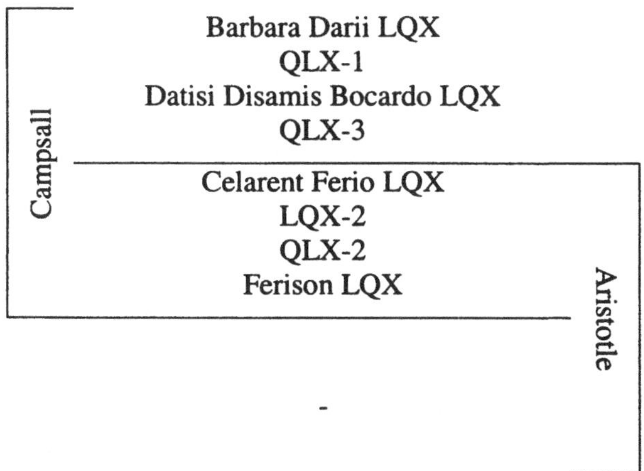

Figure 7.7 Campsall's L/X/Q system and Aristotle's system

L/Q/M combinations

Once again, the syllogisms in this layer possess a derivative validity, this time derived from the L/M layer. Because Q-propositions imply the corresponding M-propositions, any inference that is valid with an M-premise remains valid if that premise is replaced by the corresponding Q-premise. Thus, given the Campsall L/M moods listed in Table 7.1, we have the L/Q/M moods listed in Table 1.9.

X/Q/M combinations

Again, Campsall's system is maximal, and Aristotelian.

Table 7.8 Campsall's system: X/Q/M combinations

Figure	*Combinations*
1	XQM, QXM
2	XQM, QXM
	XQM, QXM

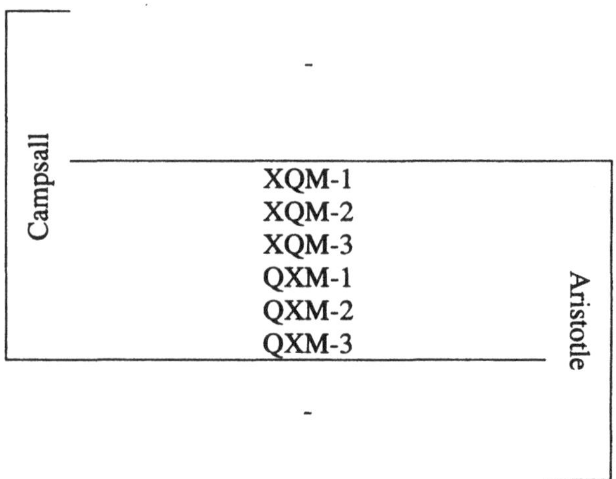

Figure 7.8 Campsall's X/Q/M system and Aristotle's system

[**Theorem 7.25a.** The XQM moods are valid in Figure 3.]
Theorem 7.27a. The XQM moods are valid in Figure 1.[49] It is noteworthy that Campsall does not think that the major in these moods need be a *simpliciter* assertoric.
[**Theorem 7.31.** The XQM moods are valid in Figure 2.] This, like all XQM moods, is valid because the XXM moods are valid.
[**Theorem 7.28a.** The QXM moods are valid in Figure 2.]
[**Theorem 7.29a.** The QXM moods are valid in Figure 1.]
[**Theorem 7.32.** The QXM moods are valid in Figure 3.] This, like the other QXM moods, is valid because the XXM moods are valid.

Summary

As Lagerlund notes,[50] Campsall's work is not easy to interpret. Accordingly, the interpretations offered in the present chapter are in many instances tentative. Nonetheless, it is clear that Campsall makes use of the basic essentialist assumption that some individuals possess some of their properties essentially, and therefore necessarily. He also assumes that there are Kinds, i.e. that some properties are such that it is necessarily the case

[49] Campsall (1968) 18.56.
[50] Lagerlund (2000) pp.58-59.

that whatever possesses them possesses them essentially. In these respects he stands in the tradition of Averröes and Kilwardby. However, Campsall is a deviant within that tradition. He treats *de esse* and *de non esse* modals differently. Regarding *de esse* modals, he is, like Abélard, an actualist, treating modal propositions as being about the very same individuals that assertoric propositions are about. But even in the case of *de esse* modals, his analysis is unconventional. His twist to actualism is to treat the predicates as well as the subjects of modal propositions as standing for actual individuals, and this has the consequence that possibility-propositions imply the corresponding assertorics. By contrast with his treatment of *de esse* modals he understands *de non esse* modals not *de rebus* but as having an irreducibly *de sensu* aspect. For example, while 'Every animal is possibly running' is about all actual animals, 'Every animal is possibly not running' is about the *per se* terms necessarily subsumed under 'animal'. And, while the first proposition settles the facts regarding running and animals, the second does not.

Idiosyncratic it may be, but Campsall's approach is also highly systematic. Everywhere except in the Q/Q layer his results can be based on the assertoric syllogistic in conjunction with modal subordination and Rule 1.3. His system includes Aristotle's in the L/M, L/X/M *simpliciter* and L/X/Q layers. In the Q/Q, L/Q/M and X/Q/M layers Campsall replicates Aristotle's results.

Chapter 8

Ockham

Notoriously, Abélard fathered a child, the boy Astralabe. He also had intellectual heirs, though it is doubtful whether we can establish, in any causal sense, that he begat them. All we can do is to remark on the congruities between his views and those of later scholars who show no evidence of direct familiarity with the Palatine's writings. Among Abélard's intellectual heirs, William Ockham stands out as preeminent. Unlike his great predecessor Ockham explicitly considers the effect of reading one or other premise or the conclusion, either as compound or as divided. In addition to that, he takes into account Aristotle's notion of a *simpliciter* assertoric, and calculates validity or invalidity on the alternative assumptions of *simpliciter* or *ut nunc* assertoric premises.

Preliminaries

Like Abélard, Ockham bases his treatment on a distinction between the compound and divided sense of modal sentences. However, his account of this distinction is quite different from Abélard's. Abélard had been uncomfortable with the notion of a propositional sense as an entity that could itself be the subject of a proposition, and as a result of this discomfort had tended to use expressions like 'as it were' [*quasi*] whenever he tries to say what the subject of a *de sensu* proposition is. Ockham would, no doubt, as a nominalist, have been equally uncomfortable in trying to designate a propositional sense as a type of entity, but he does not need to do so in order to state what the subject of a compound modal is, because he has the notion of a propositional *dictum*, i.e. an accusative and infinite phrase derived from an assertoric sentence.[1] Given this notion of a *dictum*, Ockham sees a fundamental distinction between cases where a modal proposition contains a *dictum* and cases where there is no expressed *dictum*. Propositions containing a *dictum* are according to him ambiguous. In the compound sense they mean that a mode is predicated of the proposition

[1] Ockham (1962) II ch.9:16-18.

whose *dictum* is expressed.[2] Thus, Ockham finds a nominalistically acceptable subject of which a mode can be predicated in a compound modal proposition: the subject is itself a concrete proposition, not anything abstract such as a propositional sense. A modal proposition not containing a *dictum* according to Ockham is always equivalent to a divided-sense modal.

The semantic structures that Abélard saw as underlying *de rebus* necessity-propositions are mediated by what he called natures, so that a universal negative proposition with subject '*b*' and predicate '*a*' requires that for some '*d*', $b \rightarrow D$ and $\Box\, d \mid a$ (where '*D*' stands for a nature), and a universal affirmative requires that for some '*d*', $b \rightarrow D$ and $\Box\, d \rightarrow a$. By contrast, Ockham's view is that divided universal modals simply require that $b \rightarrow a^*$ or that $b \mid a^\dagger$. He explains what he means by the divided and the compound sense of modals as follows:

> The compound sense means that the mode is predicated of the whole proposition; the divided sense means that the predicate is predicated of the subject, mediated by a verb corresponding to the mode.[3]

He illustrates with an example:

> If no ass is white ... then 'Everything white is of necessity not an ass' is true.... And the reason why this proposition is true is (as has been mentioned frequently) that ... 'Everything white is of necessity not an ass' denotes that every proposition, in which the predicate 'ass' is denied of a pronoun indicating anything that is white, is necessary. So, if there were only three whites, at that time 'Everything white is of necessity not an ass' would denote nothing other than that each of these three propositions – 'This is not an ass', 'That is not an ass', 'That is not an ass' – is necessary.... But this necessity-proposition will not be necessary, because it could have many false singulars, and consequently it will be false. And so such a necessity-proposition is sometimes true, namely when every singular is true; and is sometimes false, namely when some singular is false.[4]

This example demonstrates that Ockham's divided-sense necessity-propositions are to be taken in a weak, not a strong, sense. They need not themselves be necessary if true, but may be ephemeral necessities.

Thus, Ockham's truth-conditions for divided-sense modals are as follows.

Definition 8.1 (La). 'Every *b* is necessarily *a*': $\underline{b} \rightarrow a^*$.

---

[2] Ockham (1962) II ch.9:12-16.

[3] Ockham (1962) IIIa ch.20:14-18.

[4] Ockham (1962) IIIa ch.32:134-161.

Definition 8.2 (Le). 'No *b* is possibly *a*': $b \mid a^{\dagger}$.
Definition 8.3 (Li). 'Some *b* is necessarily *a*': $b \smile a^{*}$.
Definition 8.4 (Lo). 'Some *b* is not possibly *a*': $b \nrightarrow a^{\dagger}$.
Definition 8.5 (Ma unampliated). 'Every *b* is possibly *a*': $b \rightarrow a^{\dagger}$.
Definition 8.6 (Me unampliated). 'No *b* is necessarily *a*': $b \mid a^{*}$.
Definition 8.7 (Mi unampliated). 'Some *b* is possibly *a*': $b \smile a^{\dagger}$.
Definition 8.8 (Mo unampliated). 'Some *b* is not necessarily *a*': $b \nrightarrow a^{*}$.
Definition 8.9 (Qa, Qe unampliated). 'Every *b* may or may not be *a*': $b \rightarrow a^{\ddagger}$.
Definition 8.10 (Qi, Qo-unampliated). 'Some *b* may or may not be *a*': $b \smile a^{\ddagger}$.

Unampliated contingency-propositions imply the corresponding unampliated possibility-propositions. The truth-conditions for necessity- and possibility-propositions are similar to Abélard's, but Ockham's understanding of them differs in a significant way from Abélard's. Ockham states:

> On which account it is to be known that for the truth of such a proposition it is required that the predicate, under its own form, belongs to that (or a pronoun indicating that) for which the subject supposits, in such a way that the mode expressed in the proposition may truly be predicated of an assertoric proposition in which that very predicate is predicated of a pronoun indicating that for which the subject supposits, analogously to the way we described for past- and future-tense propositions. So, for the truth of 'Every truth is of necessity true' it is required that every proposition, in which the predicate 'true' is predicated of anything for which the subject 'truth' supposits, is necessary; in other words, that every proposition 'This is true', 'That is true', is necessary, indicating anything for which the subject supposits.[5]

By this device Ockham has in effect reduced the truth of divided modals in a certain way to that of compound modals. The divided modal 'Every *b* is necessarily *a*' is true if and only if, for all *x*, if *x* is *b*, it is necessary that *x* is *a*. This agrees with our requirement for the truth of divided modal that $\underline{b} \rightarrow a^{*}$, but what Ockham has done here is to analyse the star-operator in terms of *de dicto* necessity, in such a way that his account of necessity is more unified and simpler than Abélard's. Abélard's account rests on two basic concepts – that of a thing's *nature* and that of *de dicto* incompatibility, or strong exclusion. By contrast, Ockham's account requires only a single basic concept, namely that of strong inclusion. That

[5] Ockham (1962) II ch.10:11-23.

in which a subject is necessarily included, for Ockham, is simply that in which the subject is strongly included.

This move of Ockham's has another important aspect. So long as modal logic is based on the concept of a thing's nature, it retains a metaphysical aspect; and certainly the systems of Abélard, as well as those of Averröes and other essentialists, are open to this charge. But Ockham, by eliminating reference to things' natures, has freed modal logic from this metaphysical baggage. Or so it seems. The challenge for Ockham is to explain how the singular propositions on which divided modals rest are ever themselves necessary. He can do this readily in the example he gives ('Every truth is of necessity true') because the subjects of those singular propositions are singular terms that (so to speak) enclose a proposition within inverted commas – singular terms such as '2 + 2 = 4'. With a singular subject like this, we can read off whether the proposition to which it refers is a necessary proposition. But with a name like 'Tracy' we cannot read off from its form alone whether it is the name of a person or a cyclone, and consequently we cannot be sure whether a proposition like 'Tracy is human' is necessary or not.

We will examine Ockham's modal syllogistic in detail presently; but before that, let us look closely at Abélard's and Ockham's semantic theories, because it turns out that if they are viewed within the context of Aristotelian ontology they are not as different as they appear. Firstly, it should be observed that $\underline{b} \rightarrow a^*$ if and only if for some 'd', $\underline{b} \rightarrow D$ and $\Box D \rightarrow a$. (i) If $b \rightarrow D$ and $\Box D \rightarrow a$, then $\Box d \rightarrow a$, so (by Axioms 1.12 and 1.11) $d^* \rightarrow a^*$, and so, since $b \rightarrow d^*$, $b \rightarrow a^*$ (by Axiom 1.3). (ii) The reverse implication will also hold provided that we make the assumption that whatever falls under a term 'a^*' falls under a *per se* term that is strongly included in 'a' (Postulate 1.1). Secondly, it should be observed that $b \mid a^\dagger$ if and only if for some 'd', $b \rightarrow D$ and $\Box D \mid a^\dagger$. (i) If $b \rightarrow D$ and $\Box D \mid a^\dagger$, then $\Box d \mid a$, so (by Axioms 1.13 and 1.11) $d^* \mid a^\dagger$, and so, since $b \rightarrow d^*$, $b \mid a^\dagger$ (Axiom 1.4). (ii) The reverse implication will also hold, given the assumption that whatever excludes a term '$a^\dagger$' is included in some *per se* term that strongly excludes 'a' (Postulate 1.2).[6]

Divided modal propositions are not convertible into modal propositions, as Ockham notes:

[6] See Thom (1998).

Concerning the conversion of necessity-propositions taken in the divided sense or its equivalent, it is to be known that they are not convertible unless some change or variation in the wording is made prior to transposing the terms.[7]
But if [possibility-] propositions are taken in the divided sense or its equivalent, then they do not convert in the same way.... It is clear from the foregoing that the inference 'The seeing can be blind, so the blind can be seeing' is not valid.[8]

Ockham's main system

In presenting his system of modal syllogistic Ockham used a combinatorial methodology based on Aristotle's approach. Each chapter deals with a given combination of premise-forms in a given figure. For example chapter 20 deals with pairs of necessity-proposition in Figure 1.

Unlike Abélard Ockham wrote a *Summa*, and true to that project his treatment of modal syllogistic is systematic and highly comprehensive, covering the logic of several modalities not dealt with in a systematic way by Aristotle. Among these, the most important is (one-sided) possibility. The reason for its importance lies in the fact that possibility-propositions contradict necessity-propositions. It is an odd feature of Aristotle's modal syllogistic that it does not accommodate the contradictory of the modal propositions with which it deals; and yet Aristotle knew, from his own treatment of assertoric syllogistic, that to every syllogism there correspond two syllogisms equivalent to it by means of indirect reduction.

Ockham sees sentences containing a mode of possibility as being ambiguous, depending on whether the subject stands for those things that are or for those that can be.[9]

So if I say 'Every white thing can be a man', one sense is 'Everything that is white can be a man', and it is true in this sense if only men are white. The other sense is 'Everything that can be white can be a man', and it is false in this sense, whether only men are white or something other than a man.

It is clear from this that Ockham sometimes uses the following definitions of possibility-propositions, rather than Definitions 8.5 to 8.8.
Definition 8.11 (Ma ampliated). 'Every *b* is possibly *a*': $b^{\dagger} \rightarrow a^{\dagger}$.
Definition 8.12 (Me ampliated). 'No *b* is necessarily *a*': $b^{\dagger} \mid a^{*}$.
Definition 8.13 (Mi ampliated). 'Some *b* is possibly *a*': $b^{\dagger} \smile a^{\dagger}$.

---

[7] Ockham (1962) IIIa ch.24:49-52.
[8] Ockham (1962) IIIa ch.25:15-17, 47-48.
[9] Ockham (1962) IIIa ch.23:20-33.

Definition 8.14 (Mo ampliated). 'Some b is not necessarily a': $b^{\dagger} \not\twoheadrightarrow a^{*}$.

L/M combinations

Ockham recognizes the four first-figure LLL syllogisms, not only as valid, but as being governed by *dici de omni* when all propositions are taken in the divided sense.[10] This contrasts with Abélard's view that these moods, though valid, are not syllogistic. Ockham thus accepts the following theorem.

Theorem 8.1. The LLL moods are valid in Figure 1. Barbara: if $b \twoheadrightarrow a^{*}$ and $c \twoheadrightarrow b^{*}$ then by interpolating Axiom 1.7 we may form the chain '$c \twoheadrightarrow b^{*} \twoheadrightarrow b \twoheadrightarrow a^{*}$', which simplifies to '$c \twoheadrightarrow a^{*}$'. Celarent: if $b \mid a^{\dagger}$ and $c \twoheadrightarrow b^{*}$ then by interpolation we may form the chain '$c \twoheadrightarrow b^{*} \twoheadrightarrow b \mid a^{\dagger}$', which simplifies to '$c \mid a^{\dagger}$'. Similarly for Darii and Ferio. It appears then that, unlike Abélard, Ockham does not regard the use of interpolation as a hindrance to syllogistic validity.

Given the principles of indirect reduction, the validity of LLL in the first figure entails that of LMM in Figure 2 and MLM in Figure 3, assuming that M-propositions are contradictory to L-propositions of the opposite quantity and quality, i.e. assuming Definitions 8.5 to 8.8. Thus, Theorem 8.1 has the following corollaries:

[**Theorem 8.1a.** The LMM moods are valid in Figure 2.]

Theorem 8.1b. The MLM moods are valid in Figure 3.

Ockham does not explicitly address the question whether second-figure LMM moods are valid in the divided sense; but this is probably an oversight. He does recognize the validity of LQM moods in the second figure,[11] and of third-figure MLM moods where the conclusion's subject stands for what is.[12]

In the second figure, Ockham, like Abélard, thinks there are no valid uniform necessity-syllogisms, if the premises are taken in the divided sense.[13] He thinks that Aristotle ought not to have accepted LLL moods in the second figure. He argues with reference to Camestres LLL:

> So the Philosopher would deny that the following syllogism is good: 'Everything shining is of necessity either a heavenly body or an element, a mixed body is of necessity neither a heavenly body nor an element, so a mixed

[10] Ockham (1962) IIIa ch.20:66-78.
[11] Ockham (1962) IIIa ch.48:16-25.
[12] Ockham (1962) IIIa ch.46:8-16.
[13] Ockham (1962) IIIa ch.21:51-53.

body is of necessity not shining'. For if nothing were shining other than a heavenly body or an element, the premises would be true (in Aristotle's opinion) and the conclusion would be false, as is patently clear. That the major would indeed be true is clear because each of its singulars is true, and thus also the universal.[14]

Counter-examples are provided by Ockham to Cesare LLL where all propositions are taken in the divided sense:

For, before the incarnation the premises 'Every man of necessity is not God' and 'Every divine person of necessity is God' were true, but the conclusion 'Every divine person of necessity is not a man' was false.[15]

And to Camestres LLL:

...supposing that God suspends the action of every created agent, so that no active creature is producing anything, then the premises 'Every producer of necessity is God' and 'Every created nature of necessity is not God' are true but 'Every created nature of necessity is not producing' is false.[16]

In the third figure he accepts all LLL moods.[17]

Theorem 8.2. The LLL moods are valid in Figure 3. Datisi: if $c \rightarrow a^*$ and $c \smile b^*$ then by interpolation we may form the chain '$b \leftarrow b^* \smile c \rightarrow a^*$' which simplifies to '$b \smile a^*$'. The proof for Disamis is similar. Ferison: if $c \mid a^\dagger$ and $c \smile b^*$ then by interpolation we may form the chain '$b \leftarrow b^* \smile c \mid a^\dagger$' which simplifies to '$b \smile c \mid a^\dagger$'. The proof for Bocardo is similar. Ockham is obliged to find non-Aristotelian proofs in those cases where Aristotle relies on necessity-conversion (viz. Darapti, Felapton, Datisi, Disamis and Ferison).[18] He proves all of these by converting a premise 'Some/all c is necessarily a' to 'Something that is necessarily a is c', thus producing the first figure.[19] For Disamis and Bocardo he also offers proofs by expository syllogism.[20] Using the principles of indirect reduction, we may infer from the validity of these moods that the MLM moods in Figures 1 and 2 are valid. The first of these consequences is accepted by Ockham:

---

[14] Ockham (1962) IIIa ch.21:15-22.
[15] Ockham (1962) IIIa ch.21:33-36.
[16] Ockham (1962) IIIa ch.21:23-28.
[17] Ockham (1962) IIIa ch.22:4-32.
[18] Ockham (1962) IIIa ch.22:4-32.
[19] Ockham (1962) IIIa ch.22:5-18.
[20] Ockham (1962) IIIa ch.22:17-32.

Theorem 8.2a. The MLM moods are valid in Figure 1.[21] We saw in Chapter Two that Theophrastus recognized the validity of at least Barbara MLM, and used it to prove the validity of Bocardo LLL.[22]

Ockham does not discuss ML combinations in Figure 2, but since he accepts the second-figure QLM syllogisms,[23] he may well have accepted the second consequence:

[**Theorem 8.2b.** The MLM moods are valid in Figure 2.]

The MMM moods are invalid in Figures 1 and 3. For example, in Figure 1, if all asses are possibly white, and it happens that everything white is possibly a man, it does not follow that all asses are possibly men 'because if nothing is white except a man the premises are true and the conclusion false'.[24] Similarly in Figure 3, if every man can be white and every man can be intelligent, it does not follow that something that is intelligent can be white, 'because if no man were intelligent but an angel, the premises would be true and the conclusion false'.[25] Putting all this together, we summarize Ockham's system of divided necessity- and/or possibility-modals with unampliated subjects in Table 8.1. Absent are the Aristotelian combinations LMM-1, LLL-2 and LMM-3.

Table 8.1 Ockham's system: L/M combinations[26]

Figure	*Combinations*
1	LLL, MLM
2	LMM, MLM
3	LLL, MLM

L/X/M combinations

Abélard had recognized the validity of a number of L/X combinations via concrete examples, not general formal descriptions. The passing of two centuries would have been in vain for logical theory, if Ockham had replicated this stance. As we shall see, his treatment of these combinations is considerably more sophisticated than Abélard's.

---

[21] Ockham (1962) IIIa ch.44;69-76.

[22] See above p.22.

[23] Ockham (1962) IIIa, ch.48:16-25.

[24] Ockham (1962) IIIa, ch.23:52-56.

[25] Ockham (1962) IIIa ch.25:26-30.

[26] See Appendix III in Lagerlund (2000) p.237, Table 32.4 in Thom (1996) p.245, and Table 4b in Thom (1997) p.696 – which however is wrongly labelled 'Buridan'.

Barbara LXL is valid according to Ockham if the major is taken in the divided sense.[27] He goes on to say that the conclusion follows only in the divided, not the compound, sense, because from the premises 'Any divine person is of necessity God, a creator is a divine person' it does not follow that the proposition 'A creator is God' is necessary, though it does follow that a creator is necessarily God.

This, he says, is what must be said 'according to the truth', as opposed to what must be said 'according to the Philosopher's meaning'. The latter he expounds thus:

> ...to make this mood valid it would suffice to subsume something *per se* in a genus importing a distinct absolute thing, so that it was not a connotative or relative name. So according to the Philosopher's meaning the syllogism 'For every animal to be coloured is necessary, every man is an animal, so for every man to be coloured is necessary' is good, so that according to him whenever something absolute that is not relative or connotative is taken under the subject, the mood is always good.[28]

In other words, Ockham thinks that Aristotle holds Barbara LXL to be valid if the major is a compound necessity-proposition and the minor has a *per se* subject. His understanding of Aristotle's view owes something perhaps to Averröes. But against this view he objects that in the argument 'For every being to be in act is necessary, every man is a being, so for every man to be in act is necessary' the premises are true and the conclusion false. All the first-figure LXL are valid according to Ockham, but none of the first-figure XLL moods are valid.[29]

Theorem 8.3. The LXL moods are valid in Figure 1. Barbara: if $b \rightarrow a^*$ and $c \rightarrow b$ then we have the chain '$c \rightarrow b \rightarrow a^*$', which simplifies to '$c \rightarrow a^*$'. Celarent: if $b \mid a^\dagger$ and $c \rightarrow b$ then we have the chain '$c \rightarrow b \mid a^\dagger$' which simplifies to '$c \mid a^\dagger$'. Darii: if $b \rightarrow a^*$ and for some 'd' $c \leftarrow d \rightarrow b$ then we have the chain '$c \leftarrow d \rightarrow b \rightarrow a^*$' which simplifies to '$c \leftarrow d \rightarrow a^*$'. Ferio: if $b \mid a^\dagger$ and for some 'd' $c \leftarrow d \rightarrow b$ then we have the chain '$c \leftarrow d \rightarrow b \mid a^\dagger$' which simplifies to '$c \leftarrow d \mid a^\dagger$'.

In accordance with the principles of indirect reduction, the validity of Figure 1 LXL carries with it the validity of Figure 2 LMX and Figure 3 MXM. As noted earlier, Ockham does not consider the LM combinations of divided modals in Figure 2, but he does accepts LQX divided moods in

---

[27] Ockham (1962) IIIa ch.31:29-30.

[28] Ockham (1962) IIIa ch.31:78-85.

[29] Ockham (1962) IIIa ch.31:7-28, 95-102.

that figure;[30] and he does accept the third-figure MXM moods where the possibility-propositions are taken in the divided sense.[31] So it is safe to attribute the following two theorems to him:

[**Theorem 8.3a.** The LMX moods are valid in Figure 2.]

[**Theorem 8.3b.** The MXM moods are valid in Figure 3.]

In the second figure, Ockham says, there are no valid mixed necessity-assertoric moods with divided modals.[32] This agrees with Abélard's position.

In the third figure, Ockham, like Abélard, recognizes the validity of all six LXL moods and no XLL moods:[33]

Theorem 8.4. The LXL moods are valid in Figure 3. Datisi: if $c \rightarrow a^*$ and $c \leftarrow d \rightarrow b$ then we have the chain '$b \leftarrow d \rightarrow c \rightarrow a^*$' which simplifies to '$b \leftarrow d \rightarrow a^*$'. Disamis: if $c \leftarrow d \rightarrow a^*$ and $c \rightarrow b$ then we have the chain '$b \leftarrow c \leftarrow d \rightarrow a^*$' which simplifies to '$b \leftarrow d \rightarrow a^*$'. Ferison: if $c \mid a^\dagger$ and $c \leftarrow d \rightarrow b$ then we have the chain '$b \leftarrow d \rightarrow c \mid a^\dagger$' which simplifies to '$b \leftarrow d \mid a^\dagger$'. Bocardo: if for some 'd' $c \leftarrow d \mid a^\dagger$ and $c \rightarrow b$ then we have the chain '$b \leftarrow c \leftarrow d \mid a^\dagger$' which simplifies to '$b \leftarrow d \mid a^\dagger$'. Among the valid moods are the unAristotelian Disamis LXL and Bocardo LXL; and, unlike Abélard, Ockham gives proofs of their validity:

> But if the major is a particular affirmative, the mixture is valid. For it follows 'Some man of necessity is an animal, every man is white, so some white of necessity is an animal'. Such a syllogism reduces to the first figure by converting the major to an assertoric, converting the conclusion to an assertoric, and transposing the premises, thus: 'Every man is white, something that of necessity is an animal is a man, so something that of necessity is an animal is white', which is a syllogism governed by *dici de omni*.[34]

> But if the major is a particular negative, the discourse is valid, because it follows 'Something productive of necessity is not an ass, everything productive is a divine person, so some divine person of necessity is not an ass'. Such a syllogism is proved by an expository syllogism. If something productive of necessity is not an ass, let that be *A*, then *A* of necessity is not an ass and *A* is a divine person (by the minor proposition), so some divine person of necessity is not an ass.[35]

[30] Ockham (1962) IIIa ch.48:16-25.
[31] Ockham (1962) IIIa ch.36:18-23.
[32] Ockham (1962) IIIa ch.32:88ff.
[33] Ockham (1962) IIIa ch.33:36ff.
[34] Ockham (1962) IIIa ch.33:47-55.
[35] Ockham (1962) IIIa ch.33:56-63.

The first of these proofs can be represented as follows. If for some '*d*', $a^* \leftarrow d \rightarrow c$, and $c \rightarrow b$, then for some '*d*', $a^* \leftarrow d \rightarrow b$. Similarly for the second.

The validity of third-figure LXL moods entails that of first-figure MXM and second-figure MLX moods.

Theorem 8.4a. The MXM moods are valid in Figure 1. Working with Definitions 8.5 to 8.8, Ockham recognizes the first-figure MXM moods as valid if the possibility-propositions are taken in the divided sense with subjects standing for things that are.[36] This agrees with Abélard's results.

Theorem 8.4b. The MLX moods are valid in Figure 2. In the second figure, Ockham mentions no valid moods with only divided modals. This is probably an oversight, because he recognizes all second-figure QLX and LQX moods as valid.[37]

Putting Theorems 8.3 to 8.4 together, we summarize Ockham's system of L/X/M moods with unampliated modal propositions in Table 8.2.

Table 8.2 Ockham's system: L/X/M combinations[38]

Figure	*Combinations*
1	LXL, MXM
2	LMX, MLX
3	LXL, MXM

As can be seen by comparing Tables 8.1 and 8.2 with Tables 3.6 and 3.7, Ockham's system is identical with Abélard's.

Ockham sees first-figure XLL moods as being valid under certain conditions, namely if the major is a *simpliciter* assertoric,[39] in which case they are valid whether the minor is taken in the divided or the compound sense.[40] He explains what he means by a *simpliciter* assertoric:

> ...the predicate cannot belong to the subject at one time and be denied of it at another, but is always uniformly related to it, so that it is either always or never predicated of it. But an *ut nunc* assertoric is one in which the predicate can be truly affirmed of the subject at one time and truly denied of it at another time.[41]

---

[36] Ockham (1962) IIIa ch.34:26-30.

[37] Ockham (1962) IIIa ch.48:16-25.

[38] See Appendix III in Lagerlund (2000) p.237, Table 32.4 in Thom (1996) p.245, and Table 4b in Thom (1997) p.696 – which however is wrongly labelled 'Buridan' and omits Disamis LXL.

[39] Ockham (1962) IIIa ch.31:95-114.

[40] Ockham (1962) IIIa ch.31:140-142.

[41] Ockham (1962) IIIa ch.31:64-68.

Simpliciter assertorics are not the same as propositions predicating a superior of an inferior, where by an inferior Ockham means 'any term that cannot be in more things, but can be in fewer'.[42] He distinguishes two types of case here, depending on whether the inferior is *per se* (as 'man' is inferior to 'animal')[43] or *per accidens* (as 'white ass' is inferior to 'ass').[44] Both types of proposition are strong inclusions, but only the first type satisfies the requirements for a *simpliciter* assertoric, because only in the first case can we truly say that the predicate is always uniformly related to the subject. If the subject is a *per accidens* inferior of the predicate then the predicate can at different times be affirmed and denied of the subject. For example, 'ass' can be affirmed of 'white ass' so long as there are white asses, but if there are no white asses then no such affirmation is true; the corresponding denials are true. By contrast, if the subject is a *per se* term then no such fluctuation is possible. Given all this, and ignoring any distinction between necessity and omni-temporality, we can define *simpliciter* universal affirmative assertorics, and universal affirmatives predicating a superior of an inferior, as follows:

Definition 8.15 (Xa *simpliciter*). 'Every b is *simpliciter* a': $\Box\, B \rightarrow a$.

Definition 8.16 (Xe *simpliciter*). 'No b is *simpliciter* a': $\Box\, B \mid a$.

Definition 8.17 (Xi *simpliciter*). 'Some b is *simpliciter* a': $\Box$ for some 'd', $\Box\, b \leftarrow D$ and $\Box\, D \rightarrow a$.

Definition 8.18 (Xo *simpliciter*). 'Some b is *simpliciter* not a': for some 'd', $\Box\, b \leftarrow D$ and $\Box\, D \mid a$.

The contradictories of *simpliciter* assertorics are not defined.

Definition 8.19 (Xa superior/inferior). 'Every b is a': $\Box\, b \rightarrow a$.

Comparing Definition 8.15 with Definition 7.10 we see that Ockham's truth-conditions for affirmative *simpliciter* propositions are identical with Campsall's for propositions that are either *simpliciter* or *ut nunc* in the first way. (The situation is different for negatives, because of Campsall's general approach to negative apodeictics.) We may then attribute the following theorem to Ockham, relative to definitions 8.1 and 8.19:

Theorem 8.5. The XLL moods with *simpliciter* major are valid in Figure 1. Barbara: if $\Box\, B \rightarrow a$ and $c \rightarrow b^*$ then, applying Axioms 1.12 and 1.11 (and ignoring the *per se* status of 'b')[45] we may replace the major by '$b^* \rightarrow a^*$' and so we may form the chain '$c \rightarrow b^* \rightarrow a^*$' and simplify this to '$c \rightarrow a^*$'.

[42] Ockham (1962) IIIa ch.32:27-28.

[43] Ockham (1962) IIIa ch.33:20-23.

[44] Ockham (1962) IIIa ch.32:31-32.

[45] Relying, in other words, solely on the fact that 'a' is superior to 'b'.

Darii can be proved in a similar fashion. Celarent: if $\Box B \mid a$ and $c \rightarrow b^*$ then, applying Axioms 1.13 and 1.11 (and ignoring the *per se* status of 'b') we may replace the major by '$b^* \mid a^{\dagger}$' and so we may form the chain '$c \rightarrow b^* \mid a^{\dagger}$' which simplifies to '$c \mid a^{\dagger}$'. The proof for Ferio is similar. Essential to these proofs is the role played by Axioms 1.11 to 1.13. Ockham in effect recognizes Axiom 1.12 when he says:

> It is to be noted that if the major is a *simpliciter* assertoric, such a mixture holds because when something cannot be predicated of something at one time and denied at another, but is predicated either always or never, then the subject cannot be predicated with a modality of necessity unless the predicate is predicated of the same thing with the same modality.[46]

Following from Theorem 8.5 we have:
[**Theorem 8.5a.** The XMM moods with *simpliciter* major are valid in Figure 2.]

Table 8.3 Ockham's *simpliciter* system: L/X/M combinations

Figure	*Combinations*
1	LXL, XLL, XMM, MXM
2	LMX, XLL, XMM, MLX
3	LXL, MXM

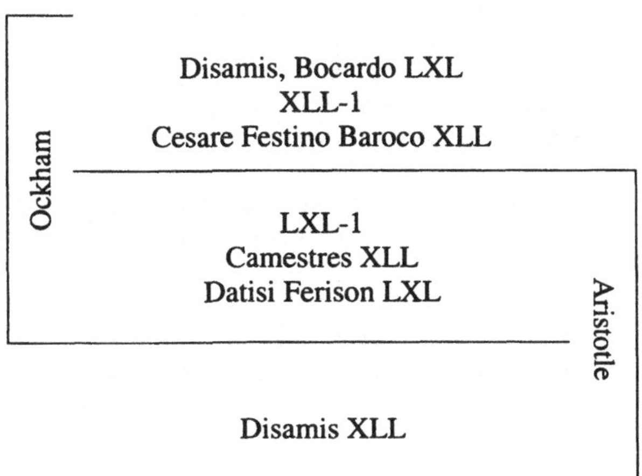

Figure 8.1 Ockham's *simpliciter* L/X/M system and Aristotle's system

[46] Ockham (1962) IIIa ch.31:110-114.

Ockham makes a noteworthy observation about a difficulty in Aristotle's treatment of the two Barbaras:

> And if it is said that the Philosopher concedes one and denies the other, so that what is said here, that one is just as valid as the other, is false – then it is to be said that the Philosopher speaks of the necessity-proposition in the divided sense; and then it is obviously clear that if the major is a necessity-proposition and the minor an assertoric, a necessity-conclusion follows whether the minor is assertoric *simpliciter* or *ut nunc*. But it is not thus if the major is assertoric, and for this reason he denies one mixture and concedes the other.[47]

Here we encounter in an acute form the problem facing an interpreter who thinks, contrary to what Aristotle says, that there are readings of both Barbara XLL and Barbara LXL that make them both valid. Ockham's way out of the difficulty carries some plausibility.

> The Philosopher does not deny this. What he denies is that, universally, from an assertoric major and a necessity-minor there follows a necessity-conclusion. He does not deny it when the major is a *simpliciter* assertoric.[48]

Using his basic interpretive strategy of always distinguishing between the compound and the divided sense of modal sentences, Ockham sees Aristotle as sometimes using one of these senses and sometimes the other:

> It is to be said that the Philosopher sometimes speaks of modals in the compound sense, and sometimes in the divided sense. Where he deals with uniform necessity-syllogisms, he speaks of them in the compound sense or its equivalent; but when he deals with mixtures in the first figure, he speaks of them in the divided sense or its equivalent.[49]

This is an ancestor of the view, prominent in Becker's work on the modal syllogistic, that Aristotle vacillates between *de re* and *de dicto* readings of modal propositions.[50]

There are other cases where *simpliciter* assertorics play a pivotal role. The second-figure XLL moods are valid, though Ockham does not mention this.

---

[47] Ockham (1962) IIIa ch.31:115-122.
[48] Ockham (1962) IIIa ch.31:142-145.
[49] Ockham (1962) IIIa ch.31:128-133.
[50] See Becker (1933).

[Theorem 8.6. The XLL moods with *simpliciter* assertorics are valid in Figure 2.] This theorem is true of all moods in Figure 2, not just the universal moods. Cesare: if $\Box\, B \mid a$ and $c \rightarrow a^{*}$ then, applying Axioms 1.15 and 1.11 we may replace the major by '$b^{\dagger} \mid a^{*}$' so that we may form the chain '$c \rightarrow a^{*} \mid b^{\dagger}$' which simplifies to '$c \mid b^{\dagger}$'. Camestres: if $\Box\, B \rightarrow a$ and $c \mid a^{\dagger}$ then, applying Axioms 1.14 and 1.11 we may replace the major by '$b^{\dagger} \rightarrow a^{\dagger}$' and so we may form the chain '$c \mid a^{\dagger} \leftarrow b^{\dagger}$' which simplifies to '$c \mid b^{\dagger}$'. Festino: if $\Box\, B \mid a$ and $c \smile a^{*}$ then, applying Axioms 1.15 and 1.11 we may replace the major by '$b^{\dagger} \mid a^{*}$' and so we may form the chain '$c \smile a^{*} \mid b^{\dagger}$'. Baroco: if $\Box\, B \rightarrow a$ and for some 'd', $c \leftarrow d \mid a^{\dagger}$, then, applying Axioms 1.14 and 1.11 we may replace the major by '$b^{\dagger} \rightarrow a^{\dagger}$' and so we may form the chain '$c \leftarrow d \mid a^{\dagger} \leftarrow b^{\dagger}$', which simplifies to '$c \leftarrow d \mid b^{\dagger}$'.

Theorem 8.6 implies the following theorem, via the principles of indirect reduction. Ockham does not mention this fact.

[Theorem 8.6a. The XMM moods with *simpliciter* assertorics are valid in Figure 1.]

Q/Q combinations

Ockham's treatment of contingency-syllogistic cannot be considered an heir to Abélard's speculations in the direct sense in which this is possible for the necessity-syllogistic, because Abélard did not discuss contingency-syllogisms. It is nonetheless formulated in the same semantic spirit that animated Abélard's modal logic.

Ockham distinguishes three senses of contingency-propositions:

> ...a contingency-propositions has three senses, depending on the way in which the subject is taken. The subject can be taken precisely for those things that are. It can also be taken equally for those things that are and for those that may or may not be. In a third way, it can be taken precisely for those things that may or may not be.[51]

What does it mean to stand equally for those things that are and those that may or may not be? It means to stand for the possible. Definitions 8.9 and 8.10 cover the first of these senses. The following pairs of definitions cover the other two senses.

Definition 8.20 (Qa, Qe – M-ampliated). 'Every b may or may not be a': $b^{\dagger} \rightarrow a^{\ddagger}$.

[51] Ockham (1962) IIIa ch.28:74-78.

Definition 8.21 (Qi, Qo – M-ampliated). ‘Some *b* may or may not be *a*’: $b^{\dagger} \smile a^{\ddagger}$.

Definition 8.22 (Qa, Qe – Q-ampliated). ‘Every *b* may or may not be *a*’: $b^{\ddagger} \rightarrow a^{\ddagger}$.

Definition 8.23 (Qi, Qo – Q-ampliated). ‘Some *b* may or may not be *a*’: $b^{\ddagger} \smile a^{\ddagger}$.

There are no valid pure contingency-syllogisms with all propositions unampliated. In the case of first-figure divided syllogisms with contingency-major, Ockham accepts QQQ moods where the major's subject stands equally for those things that are and for those that may or may not be, i.e. where the major term is ampliated to the possible.[52]

Theorem 8.7. The QQQ moods are valid in Figure 1 with M- or Q-ampliation. Barbara (Celarent): if $b^{\dagger} \rightarrow a^{\ddagger}$ and $c^{\dagger} \rightarrow b^{\ddagger}$ then by interpolation of Axiom 1.9 we may construct the chain ‘$c^{\dagger} \rightarrow b^{\ddagger} \rightarrow b^{\dagger} \rightarrow a^{\ddagger}$’, which simplifies to ‘$c^{\dagger} \rightarrow a^{\ddagger}$’. Darii (Ferio): if $b^{\dagger} \rightarrow a^{\ddagger}$ and $c^{\dagger} \leftarrow d \rightarrow b^{\ddagger}$ then by interpolation we may form the chain ‘$c^{\dagger} \leftarrow d \rightarrow b^{\ddagger} \rightarrow b^{\dagger} \rightarrow a^{\ddagger}$’ which simplifies to ‘$c^{\dagger} \leftarrow d \rightarrow a^{\ddagger}$’. This establishes validity for M-ampliation. The proofs for Q-ampliation omit the interpolation of Axiom 1.9.

Theorem 8.8. The QQQ moods are valid in Figure 3 with M- or Q-ampliation. Datisi (Ferison): if $c^{\dagger} \rightarrow a^{\ddagger}$ and $c^{\dagger} \leftarrow d \rightarrow b^{\ddagger}$, then by interpolation we may form the chain ‘$b^{\dagger} \leftarrow b^{\ddagger} \leftarrow d \rightarrow c^{\dagger} \rightarrow a^{\ddagger}$’, and thence the chain ‘$b^{\dagger} \leftarrow d \rightarrow a^{\ddagger}$’. Disamis (Bocardo): if $c^{\dagger} \leftarrow d \rightarrow a^{\ddagger}$ and $c^{\dagger} \rightarrow b^{\ddagger}$ then by interpolation we may form the chain ‘$b^{\dagger} \leftarrow b^{\ddagger} \leftarrow c^{\dagger} \leftarrow d \rightarrow a^{\ddagger}$’ and thence the chain ‘$b^{\dagger} \leftarrow d \rightarrow a^{\ddagger}$’.

We may therefore attribute to him the pure contingency-syllogisms in Table 8.4, on the assumption of Definitions 8.17 and 8.18.

Table 8.4 Ockham's Q/Q system with M- or Q-ampliation

Figure	*Combinations*
1	QQQ
2	-
3	QQQ

X/Q combinations

In the case of first-figure mixed divided syllogisms with contingency-major, Ockham accepts Aristotle's results in all cases where the major's

[52] Ockham (1962) IIIa ch.37:26-35, 43-46.

subject stands just for those things that are.[53] Thus Ockham is saying that the first-figure QXQ syllogisms are valid, indeed perfect, where the major's subject is unampliated.

Theorem 8.9. The QXQ moods are valid in Figure 1. Barbara (Celarent): if $b \rightarrow a^{\ddagger}$ and $c \rightarrow b$ then we have the chain '$c \rightarrow b \rightarrow a^{\ddagger}$', which simplifies to '$c \rightarrow a^{\ddagger}$'. Darii (Ferio): if $b \rightarrow a^{\ddagger}$ and $c \leftarrow d \rightarrow b$ then we have the chain '$c \leftarrow d \rightarrow b \rightarrow a^{\ddagger}$' which simplifies to '$c \leftarrow d \rightarrow a^{\ddagger}$'.

In the third figure he accepts Darapti, Felapton, Datisi and Ferison in the form QXQ (where the major term stands for those things that are), since these moods can be reduced to the first figure by converting the assertoric minor premise.[54]

Theorem 8.10. The QXQ moods are valid in Figure 3. Datisi (Ferison): if $c \rightarrow a^{\ddagger}$ and $c \smile b$ then $b \smile c \rightarrow a^{\ddagger}$ and so $b \smile a^{\ddagger}$. Disamis (Bocardo): if $c \smile a^{\ddagger}$ and $c \rightarrow b$ then $b \leftarrow c \smile a^{\ddagger}$ and so $b \smile a^{\ddagger}$.

Ockham proposes another theorem when the assertoric premise is *simpliciter*:

Theorem 8.11. Barbara and Darii XQQ are valid with *simpliciter* assertoric.[55] Here perhaps we see the influence of Campsall, who had identified an instance of Barbara XQQ that is valid because its premises are mutually incompatible, the major implying that every *b* is necessarily *b*, and the minor implying that some *b* is contingently *b*.[56] Ockham does not explain why he thinks these moods are valid, but perhaps he too thinks that their premises are incompatible. Now, it is true on Ockham's semantics that the minor ($\underline{c} \rightarrow b^{\ddagger}$) implies that something (some *c*) is contingently *b*, and that the major ($\Box B \rightarrow a$) requires that every *b* is necessarily *b*. But there is no incompatibility here, if some non-*b* might be contingently *b*.[57] The premises, however, will be incompatible if the *per se* status of '*b*' requires that whatever is possibly *b* is actually *b*. In that case, the major will imply that nothing is contingently *b*, because what is *b* must be *b* and what is not *b* cannot be *b*. In order to save Theorem 8.11, we may tentatively attribute to Ockham the following restricted version of Campsall's Thesis.

Theorem 8.12 (Ockham's Thesis). If '*a*' is *per se* then if $x \rightarrow a^{\dagger}$ then $x \rightarrow a$.

---

[53] Ockham (1962) IIIa ch.37:26-35, 43-46.

[54] Ockham (1962) III1 ch.39:21-25. Patterson (1995) pp.200-201.

[55] Ockham (1962) IIIa ch.37:52-55.

[56] See above p.134.

[57] Note that even if these moods are valid they are not syllogistic, and that the premises are not incompatible if the major is not *simpliciter* but merely predicates a superior of an inferior.

Aristotle had not accepted Disamis (Bocardo) QXQ, and had accepted Disamis (Bocardo) XQQ; so Ockham's system overlaps with that of Aristotle, as shown in Figure 8.2.

Table 8.5 Ockham's system: X/Q combinations

Figure	*Combinations*
1	QXQ, Barbara Darii XQQ
2	-
3	QXQ

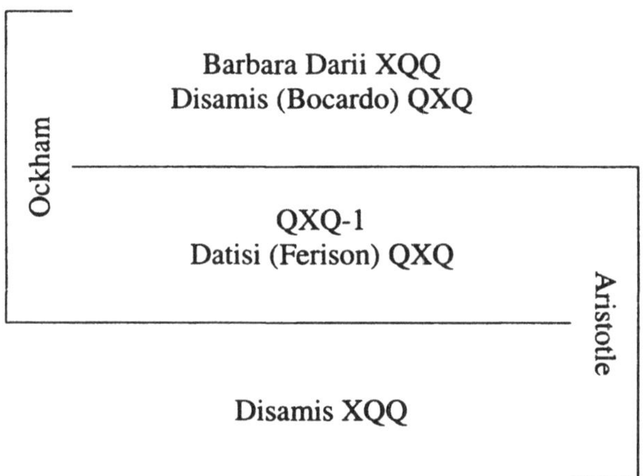

Figure 8.2 Ockham's X/Q system and Aristotle's system

L/Q combinations

Given Theorems 8.10 and 8.11, and given that necessity-propositions imply the corresponding assertorics, Ockham should accept the QLQ moods in the first and third figures. This he does.[58]

Theorem 8.13. The QLQ moods are valid in Figure 1.

Theorem 8.14. The QLQ moods are valid in Figure 3.

Aristotle (Table 1.7) had accepted Disamis LQQ, but for Ockham this mood is invalid, as it would require modal conversions.

[58] Ockham (1962) IIIa ch.47:89-104; ch.49:30-38.

Table 8.6 Ockham's system: L/Q combinations

Figure	*Combinations*
1	QLQ
2	-
3	QLQ

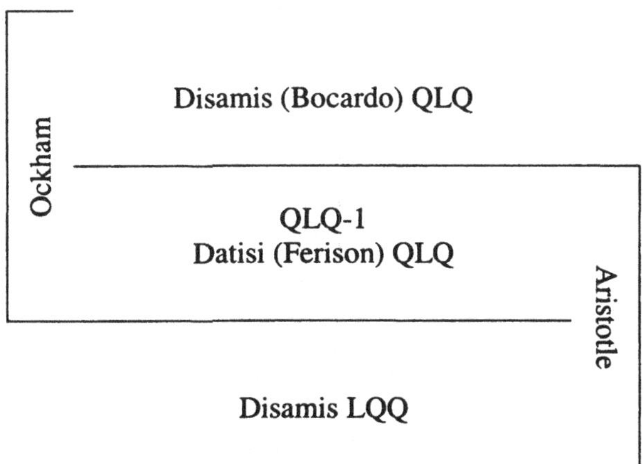

Figure 8.3 Ockham's L/Q system and Aristotle's system

L/X/Q combinations

Given that Ockham is committed to accepting the LMX and MLX moods in Figure 2 (Theorems 8.3a and 8.4b), we would expect him to accept the LQX and QLX moods in that figure. This he does.[59]

Theorem 8.15. The LQX moods are valid in Figure 2.

Theorem 8.16. The QLX moods are valid in Figure 2.

Aristotle had accepted Celarent and Ferio LQX, but his proofs depend on necessity-conversion, which is invalid for Ockham. These moods are invalid on Ockham's semantics because we cannot find a genuine middle term. Ockham's system is properly included in Aristotle's, as is shown in Figure 8.4.

[59] Ockham (1962) IIIa ch.48:16-25.

Table 8.7 Ockham's system: L/X/Q combinations

Figure	*Combinations*
1	-
2	LQX, QLX
3	-

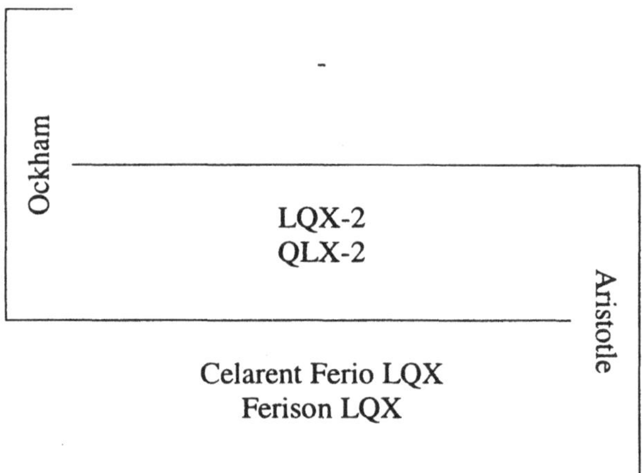

Figure 8.4 Ockham's L/X/Q system and Aristotle's system

L/Q/M combinations

Aristotle (Table 1.9) accepts the LQM and QLM moods in all figures. Given Ockham's system of L/M combinations (Table 8.1), he ought to accept QLM moods in Figure 1, LQM and QLM moods in Figure 2, and QLM moods in Figure 3 – by virtue of the fact that a Q-premise entails the corresponding M-premise. The QLM moods in Figures 1 and 3 could also be seen as derivative upon the QLQ moods in those figures.

Theorem 8.17. The LQM moods are valid in Figure 2.[60] Cesare: if $b \mid a^{\dagger}$ and $c \rightarrow a^{\ddagger}$ then, interpolating Axioms 1.9 and 1.7, we may form the chain '$c \rightarrow a^{\ddagger} \rightarrow a^{\dagger} \mid b \leftarrow b^{*}$', which simplifies to '$c \mid b^{*}$'. Camestres: if $b \rightarrow a^{*}$ and $c \rightarrow a^{\ddagger}$ then, interpolating Axioms 1.10 and 1.7, we may form the chain '$c \rightarrow a^{\ddagger} \mid a^{*} \leftarrow b \leftarrow b^{*}$', which simplifies to '$c \mid b^{*}$'. Festino and Baroco can be proved similarly.

[60] Ockham (1962) IIIa ch.48:16-25.

Theorem 8.18. The QLM moods are valid in Figure 1.[61] Barbara: if $b \rightarrow a^{\ddagger}$ and $c \rightarrow b^{*}$ then, interpolating Axioms 1.7 and 1.9, we may form the chain '$c \rightarrow b^{*} \rightarrow b \rightarrow a^{\ddagger} \rightarrow a^{\dagger}$', which simplifies to '$c \rightarrow a^{\dagger}$'. Celarent: if $b \rightarrow a^{\ddagger}$ and $c \rightarrow b^{*}$ then, interpolating Axioms 1.7 and 1.10, we may form the chain '$c \rightarrow b^{*} \rightarrow b \rightarrow a^{\ddagger} \mid a^{*}$', which simplifies to '$c \mid a^{*}$'. Darii and Ferio can be proved similarly.
Theorem 8.19. The QLM moods are valid in Figure 2.[62] Cesare: if $b \rightarrow a^{\ddagger}$ and $c \rightarrow a^{*}$ then, interpolating Axioms 1.10 and 1.7, we may form the chain '$c \rightarrow a^{*} \mid a^{\ddagger} \leftarrow b \leftarrow b^{*}$', which simplifies to '$c \mid b^{*}$'.Camestres: if $b \rightarrow a^{\ddagger}$ and $c \mid a^{\dagger}$ then, interpolating Axioms 1.9 and 1.7, we may form the chain '$c \mid a^{\dagger} \leftarrow a^{\ddagger} \leftarrow b \leftarrow b^{*}$', which simplifies to '$c \mid b^{*}$'. Similarly for Festino and Baroco.
Theorem 8.20. The QLM moods are valid in Figure 3.[63]

Table 8.8 Ockham's system: L/Q/M combinations

Figure	*Combinations*
1	QLM
2	LQM, QLM
3	QLM

Ockham's and Aristotle's results are compared in Figure 8.5.

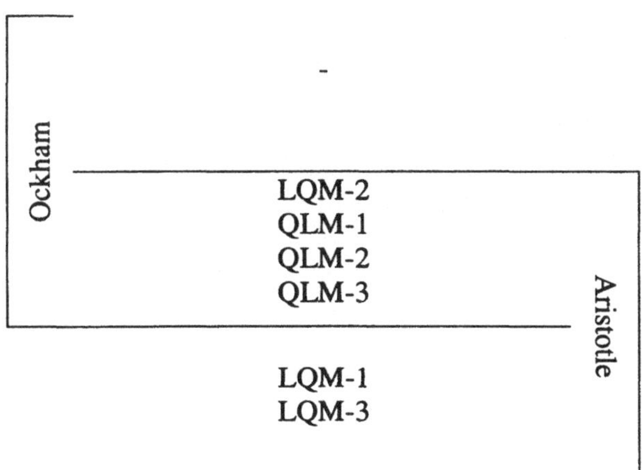

Figure 8.5 Ockham's L/Q/M system and Aristotle's system

[61] Ockham (1962) IIIa ch.47:100-102.
[62] Ockham (1962) IIIa ch.48:16-25.
[63] Ockham (1962) IIIa ch.46:14-19.

The LQM moods are invalid in Figure 1. Ockham gives the counter-example 'Every knower of necessity is God, every man can be a knower, so every man can be God'.[64] The LQM moods are also invalid in Figure 3. Ockham gives the counter-example 'Every divine person of necessity is God, some divine person can be a man, so someone who is a man can be God'.[65] These results are displayed in Table 8.8.

X/Q/M combinations

Aristotle (Table 1.10) had accepted the XQM moods in Figure 1, but for Ockham these moods are invalid unless the assertoric is *simpliciter*. Ockham uses the counter-example 'Everything white is a man, every horse may or may not be white' to show that a contingency-conclusion does not follow;[66] and it is equally obvious that there is no possibility-conclusion. Aristotle had accepted Cesare and Festino XQM. Ockham states that Aristotle is to be understood as meaning a *simpliciter* assertoric here.[67] Without that rider, these moods are invalid for Ockham: there is no genuine middle term since the major's predicate in unmodalized and the minor's is Q-modalized. Aristotle had accepted the QXM moods in figures 1 and 3. Ockham is committed to agreeing, given his acceptance of the QXQ moods in those figure; and this he does.

Theorem 8.21. The QXM moods are valid in Figure 1.

Theorem 8.22. The QXM moods are valid in Figure 3.[68]

Aristotle had accepted Datisi and Ferison XQM, but Ockham cannot agree because the minor term in its premise is Q-modalized and would have to be unmodalized in the conclusion. Thus he accepts the syllogisms listed in Table 8.9.

Table 8.9 Ockham's system: X/Q/M combinations

Figure	*Combinations*
1	QXM
2	-
3	QXM

[64] Ockham (1962) IIIa ch.44:64-67.
[65] Ockham (1962) IIIa ch.46:11-13.
[66] Ockham (1962) IIIa ch.37:47-49.
[67] Ockham (1962) IIIa ch.38:37-40.
[68] Ockham (1962) IIIa ch.39:5-14.

Ockham notices that there are several cases where an invalid combination becomes valid when an ordinary assertoric is replaced by a *simpliciter* assertoric. He says that Celarent and Ferio XQM are valid with *simpliciter* assertoric.[69] As noted earlier, he thinks that Barbara and Darii XQQ are valid with *simpliciter* assertoric; and, while this depends on Ockham's Thesis, the corresponding XQM moods are valid independently of Ockham's Thesis. So we can prove the more general Theorem 8.23.

Table 8.10 Ockham's *simpliciter* system: X/Q/M combinations

Figure	*Combinations*
1	XQM, QXM
2	Cesare Festino XQM
3	QXM

If we compare these results with Aristotle's (Table 1.10), we find that Ockham's system is a proper part of Aristotle's. See Figure 8.6.

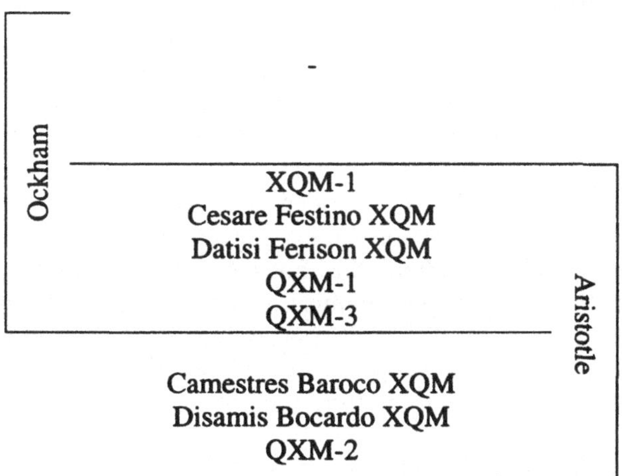

Figure 8.6 Ockham's *simpliciter* X/Q/M system and Aristotle's system

Theorem 8.23. The XQM moods are valid in Figure 1 with *simpliciter* assertoric. Barbara: if $\Box B \rightarrow A$ and $c \rightarrow b^{\ddagger}$ then, applying Axioms 1.14 and 1.11 we may replace the major with '$b^{\dagger} \rightarrow a^{\dagger}$' (ignoring the *per se* status of

[69] Ockham (1962) IIIa ch.37:55-61.

the terms), and interpolating Axiom 1.9, we may form the chain '$c \rightarrow b^{\ddagger} \rightarrow b^{\dagger} \rightarrow a^{\dagger}$', which simplifies to '$c \rightarrow a^{\dagger}$'. Celarent: if $\Box B \mid A$ and $c \rightarrow b^{\ddagger}$ then, applying Axioms 1.15 and 1.11 we may replace the major by '$b^{\dagger} \mid a^{*}$', and interpolating Axiom 1.9, we may form the chain '$c \rightarrow b^{\ddagger} \rightarrow b^{\dagger} \mid a^{*}$', which simplifies to '$c \mid a^{*}$'. Darii and Ferio can be proved in a similar fashion.

Theorem 8.24. Cesare and Festino XQM are valid with *simpliciter* assertoric.[70] Cesare: if $\Box B \mid a$ and $c \rightarrow a^{\ddagger}$ then, applying Axioms 1.13 and 1.11 we may replace the major by '$b^{*} \mid a^{\dagger}$', and interpolating Axiom 1.9, we may form the chain '$c \rightarrow a^{\ddagger} \rightarrow a^{\dagger} \mid b^{*}$', which simplifies to '$c \mid b^{*}$'. Festino can be proved similarly.

Alternate systems

Alongside his main system Ockham sketches a number of other systems, based on alternate readings of modal propositions. Of these, the two most important are the compound and ampliated readings.

Compound modals

Ockham explores different cases in which one or more of the propositions is taken in the compound sense. Compound propositions of a given modality are true, according to him, just in case that modality is truly predicated of the corresponding propositional *dictum*, so a compound necessity-proposition is true provided that the corresponding *dictum* is necessary, and so on. The principles governing the logic of compound modals are that (i) 'when the antecedent is necessary so is the consequent',[71] (ii) 'when one premise is necessary and the other possible, the conclusion is possible',[72] and (iii) a contingency-proposition implies the corresponding possibility-proposition.[73] Given these principles, we can deduce the following theorems from the validity of the standard assertoric syllogisms.

Theorem 8.25. The LLL compound moods are valid in all figures.

Theorem 8.26. The MLM compound moods are valid in all figures.[74]

Theorem 8.27. The LMM compound moods are valid in all figures.[75]

---

[70] Ockham (1962) IIIa ch.38:34-40.

[71] Ockham (1962) IIIa ch.20:38-50; ch.21:1-13; ch.22:1-4.

[72] Ockham (1962) IIIa ch.44:11-12.

[73] Ockham (1962) IIIa ch.47:23.

[74] Ockham (1962) IIIa ch.44:15-18; ch.45:5-8; ch.46:1-2.

[75] Ockham (1962) IIIa ch.44:6-8; ch.45:5-8; ch.46:1-2.

Theorem 8.28. The QLM compound moods are valid in all figures.[76]
Theorem 8.29. The LQM compound moods are valid in all figures.[77]

Table 8.11 Ockham's system of compound modals: L/Q/M combinations

Figure	*Combinations*
1	LLL, LQM, LMM, QLM, MLM
2	LLL, LQM, LMM, QLM, MLM
3	LLL, LQM, LMM, QLM, MLM

Ockham saw that this repertoire could be expanded to include LX and XL combinations if assertorics were *simpliciter* (or at least if they expressed superior/inferior predications). Principle (i) above implies the validity of all LXL and XLL moods.

Theorem 8.30. The LXL compound moods with strong assertorics are valid in all figures.[78]

Theorem 8.31. The XLL compound moods with strong assertorics are valid in all figures.[79]

As corollaries of Theorems 8.30 and 8.31 we have:

Theorem 8.30a. The MXM compound moods with strong assertorics are valid in all figures.

Theorem 8.31a. The XMM compound moods with strong assertorics are valid in all figures.

Consequent upon these last two are the following theorems.

Theorem 8.32. The QXM compound moods with strong assertorics are valid in all figures.[80]

Theorem 8.33. The XQM compound moods with strong assertorics are valid in all figures.[81]

Ockham's system of compound modals combined with strong assertorics is summarized in Table 8.12.

---

[76] Ockham (1962) IIIa ch.47:30-40; ch.49:1-11.
[77] Ockham (1962) IIIa ch.47:4-18; ch.49:1-11.
[78] Ockham(1962) IIIa ch.31:59-61; ch.32:19-42; ch.33:12-28.
[79] Ockham (1962) IIIa ch.31:110-114; ch.33:12.
[80] Ockham (1962) IIIa ch.37:4-13; ch.34:15-16.
[81] Ockham (1962) IIIa ch.37:4-13; ch.34:15-16.

Table 8.12 Ockham's system of compound modals with strong assertorics: L/X/Q/M combinations

Figure	*Combinations*
1	LXL, XLL, XQM, XMM, QXM, MXM
2	LXL, XLL, XQM, XMM, QXM, MXM
3	LXL, XLL, XQM, XMM, QXM, MXM

Ampliated modals

In the summative spirit that pervades his work, Ockham explores the effects that ampliation has on modal syllogistic. These explorations involve temporarily adopting definitions that ampliate the subject of possibility- or contingency-propositions to the possible, and of ampliating the subject of contingency-propositions to the contingent. Remarkably (as Lagerlund notes),[82] they do not include an exploration of the effects of ampliating the subjects of necessity-propositions. To that extent, Ockham's study of ampliated modals is radically incomplete.

Using Definitions 8.11 to 8.14, Ockham recognizes a number of possibility-syllogisms. In the first figure, with all propositions taken in the divided sense, he recognizes MMM syllogisms in each of the four familiar moods, when the major's subject stands for those things that can be.

Theorem 8.34. The MMM moods are valid in Figure 1.[83] Barbara: if $b^{\dagger} \rightarrow a^{\dagger}$ and $c^{\dagger} \rightarrow b^{\dagger}$ then $c^{\dagger} \rightarrow a^{\dagger}$. Celarent: if $b^{\dagger} \mid a^{\dagger}$ and $c^{\dagger} \rightarrow b^{\dagger}$ then $c^{\dagger} \mid a^{\dagger}$. By indirect reduction, Theorem 8.34 should imply two corollaries about MLL moods in Figure 2 and LML moods in Figure 3; but, as we have seen, Ockham shies away from necessity-propositions with ampliated subjects, and so he did not formulate these corollaries. We will find them in Buridan's work, discussed in the next chapter.

In the third figure, all uniform possibility-moods are valid with all propositions taken in the divided sense, provided that the conclusion's subject is taken for those things that can be.[84]

Theorem 8.35. The MMM moods are valid in Figure 3. Ferison: if $c^{\dagger} \mid a^{*}$ and for some 'd', $c^{\dagger} \leftarrow d \rightarrow b^{\dagger}$, then we have the chain '$b^{\dagger} \leftarrow d \rightarrow c^{\dagger} \mid a^{*}$', which simplifies to '$b^{\dagger} \mid a^{*}$'. And again there should be two corollaries, this time about LML moods in Figure 1 and LLM moods in Figure 2, but Ockham cannot or will not consider necessity-propositions with ampliated subjects. We will find these corollaries in Buridan.

[82] Lagerlund (2000) p.91.
[83] Ockham (1962) IIIa ch.23:36-40. See Theorem 9.4.
[84] Ockham (1962) IIIa ch.25: 1-3. See Theorem 9.5.

Further complications ensue. Ockham considers the effects of combining compound-sense and divided-sense propositions in the one syllogism, with or without ampliation, with or without *simpliciter* assertorics. But one is left wondering what is the point of it all. It cannot be that he is struggling to find a satisfactory interpretation of Aristotle: he tells us himself that Aristotle sometimes means things in the compound sense (for necessity-conversion, and for the pure necessity-moods)[85] and sometimes in the divided sense (for the mixed necessity-moods).[86] On the other hand, even though he occasionally contrasts what is the case according to Aristotle and according to the truth, his aim in weaving these complications does not seem be to construct a unified theory of modal logic according to the truth.

Summary

Within his main system Ockham treats modal propositions in an actualist fashion, i.e. with no ampliation. He makes use of the basic essentialist assumption that some individuals possess some of their properties essentially, and therefore necessarily. An interesting aspect of his thought is that he wants to treat these singular necessities as compound modal propositions. Except in his discussions of *simpliciter* assertorics, Ockham does not use the stronger essentialist assumption that there are Kinds, i.e. that some properties are such that it is necessarily the case that whatever possesses them possesses them essentially.

Ockham's overall approach is minimalist. He has two systems, one in which assertorics are taken in the ordinary fashion, the other in which they are read *simpliciter.* The ordinary system is as follows. In the L/M and L/X/M layers, given that L-propositions imply X-propositions, we may take the LXL moods in Figures 1 and 3 (or their equivalents) as axiomatic, along with the QXQ moods in Figures 1 and 3. His *simpliciter* system expands this axiomatic base to include the XLL moods in Figures 1 and 2 (or their equivalents), the XQM moods in Figure 1, Cesare and Festino XQM. The QQQ moods (with ampliated propositions) may be added to either system.

Abélard pursued his basic *de rebus* approach to modal logic at an informal level. It was formalized by Ockham – perhaps excessively so, in the scholastic manner. His treatment is nonetheless incomplete, as becomes clear to any reader of Buridan.

[85] Ockham (1962) II ch.24:25-28; IIIa ch.31:130-132.
[86] Ockham (1962) IIIa ch.31:132-133.

Chapter 9

Buridan

With Buridan (c.1300 – c.1358), modal syllogistic reaches its highest point in the Middle Ages, both in terms of completeness and elegance. Lagerlund observes that it was Buridan who gave modal syllogistic the form it would retain for the rest of the Middle Ages, and that his achievement in this field 'surpassed most things done in the history of logic since Aristotle completed the *Prior Analytics*'.[1] In this chapter we will look at his modal logic as expounded in Buridan's *Tractatus de Consequentiis* and his *Summulae de Dialectica*.

Preliminaries

Buridan clearly distinguishes between necessity-propositions and propositions that are necessary, and similarly for the other modes.[2] He notes, for example, that though 'A man is of necessity running' is a necessity-proposition it is by no means a necessary proposition.

Buridan, like Ockham and Abélard before him, draws a distinction between compound and divided modals, but there are important differences in the ways these three logicians draw the distinction. For Buridan, whether a proposition is compound or divided depends on what its subject and predicate are. If the subject or predicate is a mode (such as 'necessary' or 'possible' or 'contingent') then the proposition is compound.[3] If the mode is a determination of the copula then the proposition is divided, and in this case its subject and predicate will be the same as the subject and predicate of the corresponding non-modal proposition.[4] How does this account compare with the accounts given by earlier modal logicians? On the one hand, Buridan's account of compound modals is more elegant than Ockham's, lacking the ascription of ambiguity to propositions containing a *dictum*; Buridan cuts straight to the chase, treating propositions in which a mode is predicated of a *dictum* as compound. On the other hand, Buridan's

[1] Lagerlund (2000) p.130.
[2] Buridan (1985) 2.1.3.
[3] Buridan (1985) 2.2.2.
[4] Buridan (1985) 2.2.5.

syntactic account of divided modals shares two features with Abélard's account of *de rebus* modals – the requirements that the mode be a determination of the copula, and that divided modals share their subjects and predicates with the assertorics from which they derive.

Buridan makes the interesting observation that if any sentence '*b* is *a*' is possible, or true or contingent or necessary, then every such sentence is possible etc.:

> The reason for this is that for all such sentences '*b* is *a*', one signifies whatever and however the other signifies. Therefore, if it is the case as one signifies it is the case as the other signifies....[5]

Like Avicenna, Buridan believed that the subject of a modal proposition (at least a divisive modal) should be ampliated to the possible,[6] and he illustrated this doctrine with a homely example – that of boiling water:

> In this way it is true that air can come from water, although this is not true for any air which exists.[7]

In principle, a decision to ampliate the subject of universal affirmative necessity-propositions leaves two options open. Those propositions could be taken in either a strong or a weak sense – either stating $\Box\, b^{\dagger} \rightarrow a^{*}$ or stating $b^{\dagger} \rightarrow a^{*}$. If we look at Buridan's examples, we would think that he means these propositions to be taken in the strong sense. However, if we look at what is logically required then we will be satisfied with the weak reading, as is expressed in definitions 9.1 to 9.4.

Definition 9.1 (La). 'Every *b* is necessarily *a*': $\underline{b^{\dagger}} \rightarrow a^{*}$.

Definition 9.2 (Le). 'No *b* is possibly *a*': $b^{\dagger} \mid a^{\dagger}$.

Definition 9.3 (Li). 'Some *b* is necessarily *a*': $b^{\dagger} \smile a^{*}$.

Definition 9.4 (Lo). 'Some *b* is not possibly *a*': $b^{\dagger} \nleftrightarrow a^{\dagger}$.

As can be seen from these definitions, the universal negative necessity-proposition is convertible, but the particular affirmative is not. One unusual feature of the definitions is that the standard laws of modal subordination hold only for universal negative necessity-propositions. Affirmative or particular necessity-propositions with subject '*b*' imply assertorics about possible *b*s, but not about actual *b*s. Along with Definitions 9.1 to 9.4 we will need analogous definitions of possibility- and contingency-propositions, as follows.

---

[5] Buridan (1985) 2.7.7.

[6] See Lagerlund (2000) pp.136-142.

[7] Buridan (1985) 2.4.1.

Definition 9.5 (Ma). 'Every *b* is possibly *a*': for no '*d*', $b^{\dagger} \rightarrow a^{\dagger}$.
Definition 9.6 (Me). 'No *b* is necessarily *a*': for no '*d*', $b^{\dagger} \mid a^{*}$.
Definition 9.7 (Mi). 'Some *b* is possibly *a*': $b^{\dagger} \smile a^{\dagger}$.
Definition 9.8 (Mo). 'Some *b* is not necessarily *a*': $b^{\dagger} \nrightarrow a^{*}$.
Definition 9.9 (Qa, Qe). 'Every *b* may or may not be *a*': $b^{\dagger} \rightarrow a^{\ddagger}$.
Definition 9.10 (Qi, Qo). 'Some *b* may or may not be *a*': $b^{\dagger} \smile a^{\ddagger}$.

Contingency-propositions imply the corresponding possibility-propositions.

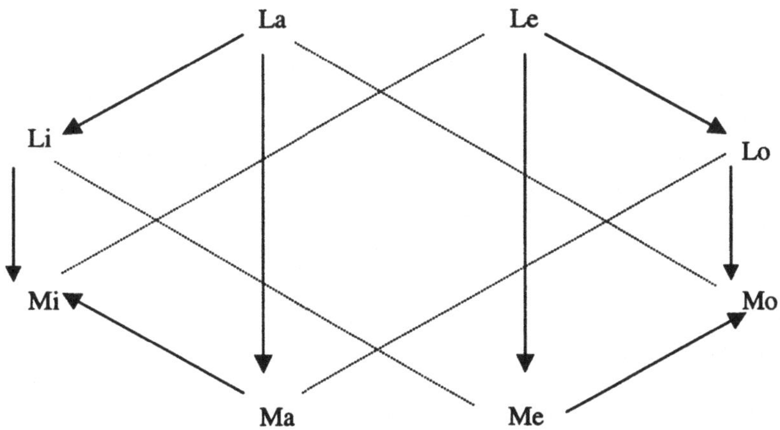

Figure 9.1 Modal octagon of opposition

Hughes uses the phrase 'the octagon of opposition' to describe the diagram in Buridan that records the logical relations between quantified modal propositions.[8] We saw that Abélard achieved the two halves of this diagram (though imperfectly). Buridan's achievement was to put the two halves together. In Figure 9.1, for the sake of perspicuity I have included only relations of contradiction and non-derivative relations of implication. The other relations usually shown in these diagrams are derivative: a contrary is what implies a contradictory, and a sub-contrary is what is implied by a contradictory.

Buridan considers the Aristotelian rule that nothing impossible follows from an assertoric proposition put forward as possible. He says that this rule should be interpreted as stating that whatever follows from the *dictum*

[8] Hughes (1989) pp.98-99. See also Lagerlund (2000) p.246.

of an affirmative possibility-proposition is possible.[9] He adds the observation that a divided possibility-proposition does not imply the possibility of the corresponding assertoric, giving the counter-examples 'Every star in the Zodiac is possibly shining above our hemisphere, but it is not possible that every star in the Zodiac is shining above our hemisphere' and 'A white thing can be black, but it is not possible that a white thing is black'.[10]

In the case of possibility-propositions, he states that a compound particular affirmative implies the corresponding divided proposition.[11] In other words, if it is possible that some *b* is *a* then some possible *b* is possibly *a*. But the inference does not work for universal propositions: if it is possible that everything running is a horse, it does not follow that everything possibly running is possibly a horse 'since perhaps an ass which cannot be a horse is running'.[12] In the case of necessity-propositions, a divided universal negative implies the corresponding compound, for example, 'No possible *b* can be *a*' implies that 'Some *b* is *a*' is impossible.[13] These results should be compared with those of Abélard and Ockham.

Buridan's main system

Buridan's method of presenting his system of modal syllogistic differs from Ockham's. Whereas Ockham divides up his chapters according to given combinations of premise-forms in a given figure, Buridan looks for broader ways of grouping his data. Further, he uses two different methodologies in the *Summulae* and the *Consequentiis*. In the *Summulae* he groups together all rules dealing with a given combination of premise-forms, no matter what their figure; for example in *Summulae* 5.7.3 we find six rules about combinations of a necessity-premise with an assertoric. In the *Consequentiis* the major principle of division is the figure of the premises; thus in Theorem IV-4 we find a single rule governing LL, MM, LM and ML premises in the first figure. In both works Buridan is seeking a higher level of generality than is evident in Ockham's presentation.

---

[9] Buridan (1985) 2.7.30; 2.7.27.
[10] Buridan (1985) 2.7.31-33.
[11] Buridan (1985) Theorem II-17.
[12] Buridan (1985) 2.7.34-35.
[13] Buridan (1985) Theorem II-18.

L/M combinations

Much of the core of Buridan's modal syllogistic can already be found in Avicenna, whose acceptance of 'substantial' first-figure moods in LLL, MMM, LML and MLM, second-figure moods in LLL, LML and MLL, and third-figure moods in LLL, MMM, LML and MLM, is mirrored by Buridan's acceptance of the same moods for divided propositions.[14]

Theorem 9.1. The LLL moods are valid in Figure 1.[15]
Theorem 9.2. The LLL moods are valid in Figure 2.[16]
Theorem 9.3. The LLL moods are valid in Figure 3.[17]

Applying the principles of indirect reduction, we can infer the following corollaries:

[**Theorem 9.1a.** The LMM moods are valid in Figure 2.]
Theorem 9.1b. The MLM moods are valid in Figure 3.[18]
[**Theorem 9.2a.** The LMM moods are valid in Figure 1.]
[**Theorem 9.2b.** The LMM moods are valid in Figure 3.]
Theorem 9.3a. The MLM moods are valid in Figure 1.[19]
[**Theorem 9.3b.** The MLM moods are valid in Figure 2.]

Table 9.1 Buridan's system: L/M combinations[20]

Figure	*Combinations*
1	LLL, LML, LMM, MLM, MMM
2	LLL, LML, LMM, MLL, MLM
3	LLL, LML, LMM, MLM, MMM

[14] Buridan (1985) Theorems IV-4, IV-5 and IV-6. King (p.82) remarks that in Theorem IV-5 'no conclusion at all seems to be entailed by the premises'. But he is mistaken, having fallen into the trap identified by Hughes (p.98): 'a universal negative *de necessario* means "Every *a* is necessarily not *b*", and not "No *a* is necessarily *b*"...'.
[15] Buridan (1985) Theorem IV-4.
[16] Buridan (1985) Theorem IV-5.
[17] Buridan (1985) Theorem IV-6.
[18] Buridan (1985) Theorem IV-6.
[19] Buridan (1985) Theorem IV-4.
[20] See Appendix V in Lagerlund (2000) p.239, Table 32.3 in Thom (1996) p.241, and Table 4a in Thom (1997) p.696 – which however is wrongly labelled 'Ockham'.

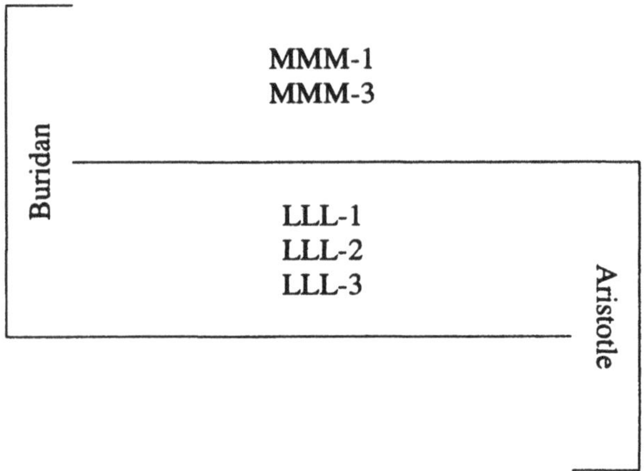

Figure 9.2 Buridan's L/M system and Aristotle's system

Buridan also states the following theorems.
Theorem 9.4. The MMM moods are valid in Figure 1.[21]
Theorem 9.5. The MMM moods are valid in Figure 3.[22]
Theorem 9.4a. The MLL moods are valid in Figure 2.[23]
Theorem 9.4b. The LML moods are valid in Figure 3.[24]
Theorem 9.5a. The LML moods are valid in Figure 1.[25]
Theorem 9.5b. The LML moods are valid in Figure 2.[26]

We may attribute to Buridan the moods listed in Table 9.1. These correspond exactly with Avicenna's L/M moods (Table 4.1). Figure 9.2 displays the relation between Buridan's system of L/M syllogisms and that of Aristotle.

L/X/M combinations

When it comes to mixed assertoric-apodeictic moods, Avicenna and Buridan agree in accepting Darii, Ferio, Festino, Datisi and Ferison LXL.

[21] Buridan (1985) Theorem IV-4.
[22] Buridan (1985) Theorem IV-6.
[23] Buridan (1985) Theorem IV-5.
[24] Buridan (1985) Theorem IV-6.
[25] Buridan (1985) Theorem IV-4.
[26] Buridan (1985) Theorem IV-5.

Theorem 9.6. Darii and Ferio LXL are valid.[27] Darii: if $b^\dagger \rightarrow a^*$ and $c \smile b$ then by interpolation we may form the chain '$c^\dagger \leftarrow c \smile b \rightarrow b^\dagger \rightarrow a^*$' which simplifies to $c^\dagger \smile a^*$'. Ferio: if $b^\dagger \mid a^\dagger$ and $c \smile b$ then by interpolation we may form the chain '$c^\dagger \leftarrow c \smile b \rightarrow b^\dagger \mid a^\dagger$' which simplifies to $c^\dagger \nrightarrow a^\dagger$'.

Theorem 9.7. Festino LXL is valid:[28] if $b^\dagger \mid a^\dagger$ and $c \smile a$ then by interpolation we may form the chain '$c^\dagger \leftarrow c \smile a \rightarrow a^\dagger \mid b^\dagger$' which simplifies '$c^\dagger \nrightarrow b^\dagger$'.

Theorem 9.8. Datisi and Ferison LXL valid.[29] Datisi: if $c^\dagger \rightarrow a^*$ and $c \smile b$ then by interpolation we may form the chain '$b^\dagger \leftarrow b \smile c \rightarrow c^\dagger \rightarrow a^*$' which simplifies to '$b^\dagger \smile a^*$'. Ferison: if $c^\dagger \mid a^\dagger$ and $c \smile b$ then we may form the chain '$b^\dagger \leftarrow b \smile c \rightarrow c^\dagger \mid a^\dagger$' which simplifies to '$b^\dagger \nrightarrow a^\dagger$'.

From each of these we can derive two corollaries, equivalent by indirect reduction. These are:

[**Theorem 9.6a.** Cesare and Camestres LMX are valid.]
Theorem 9.6b. Datisi and Ferison MXM are valid.[30]
[**Theorem 9.7a.** Celarent LMX is valid.]
Theorem 9.7b. Disamis XMM is valid.[31]
Theorem 9.8a. Darii and Ferio MXM are valid.[32]
[**Theorem 9.8b.** Cesare and Camestres MLX are valid.]

Avicenna and Buridan disagree with one another about some mixed LX and XL moods. Because Buridan analyses assertorics in the conventional way, as compared with Avicenna's unusual readings, the two systems differ in six respects.

(1) Avicenna's unconventional reading of assertorics as possibility-propositions safeguards the principles of modal subordination for universal affirmatives: if $b^\dagger \rightarrow a^*$ then $b^\dagger \rightarrow a^\dagger$, and similarly for the other apodeictic propositions. But Buridan's conventional reading of assertorics implies rejection of modal subordination in the case of particular or affirmative propositions:

> ...this does not follow, 'Some planetary body shining on our hemisphere is necessarily not the Sun; therefore, some planetary body shining on our hemisphere is not the Sun', because if we were to posit the case that only the Sun of all planetary bodies is now shining on our hemisphere, then the first

---

[27] Buridan (1985) Theorem IV-16.
[28] Buridan (1985) Theorem IV-17.
[29] Buridan (1985) Theorem IV-18.
[30] Buridan (1985) Theorem IV-13.
[31] Buridan (1985) Theorem IV-13.
[32] Buridan (1985) Theorem IV-10.

> claim... would be true... and the second false. The reason for this is that the subject in the sentence *de necessario* supposits for things other than the Sun, due to its ampliation, and in the assertoric sentence the subject does not supposit for other things.[33]

(2) Buridan holds that XM premises in the first figure are unproductive,[34] and this is correct on his readings: if $b \rightarrow a$ and $c^{\dagger} \rightarrow b^{\dagger}$ then nothing follows because there is no genuine middle term. (He points out that if everything running is a horse, and every man can run, it does not follow that every man can be a horse.) By contrast, on Avicenna's readings there is a conclusion: if $b^{\dagger} \rightarrow a^{\dagger}$ and $c^{\dagger} \rightarrow b^{\dagger}$ then $c^{\dagger} \rightarrow a^{\dagger}$.

(3) Buridan rejects Disamis and Bocardo MXM, and Datisi and Ferison XMM,[35] and this is correct on his readings: for instance if $c^{\dagger} \smile a^{\dagger}$ and $c \rightarrow b$ (Disamis MX) then there is no genuine middle term. (To demonstrate invalidity Buridan offers the terms 'Sun', 'Moon' and – as putative middle – 'planetary body shining above our hemisphere'.) But for Avicenna these inferences are valid: for instance if $c^{\dagger} \smile a^{\dagger}$ and $c^{\dagger} \rightarrow b^{\dagger}$ then $b^{\dagger} \smile a^{\dagger}$.

(4) Buridan rightly rejects Barbara and Celarent LXL:[36] for instance, if $b^{\dagger} \rightarrow a^{*}$ and $c \rightarrow b$, nothing follows. But for Avicenna '$b^{\dagger}$' is a genuine middle term here.

(5) Buridan rightly rejects Cesare, Camestres and Baroco LXL, and all second-figure XLL moods,[37] again for lack of a genuine middle term. (By way of counter-example to Camestres XLL, Buridan says, 'let us posit that every planetary body shining above our hemisphere is the Sun, and every Moon necessarily is not the Sun. It does not follow that therefore the Moon is necessarily not a planetary body shining above our hemisphere'.) On Avicenna's readings, however, there is a genuine middle.

(6) Buridan rightly rejects Disamis and Bocardo LXL,[38] which suffer from the same defect, whereas for Avicenna these moods are valid. (By way of counter-example to Disamis LXL Buridan says 'posit that the Moon is now shining, and then this does not follow: Something shining is necessarily the Moon, and everything shining is different from the Moon; therefore, something other than the Moon is necessarily the Moon'.)

---

[33] Buridan (1985) 2.6.16.
[34] Buridan (1985) Theorem IV-10.
[35] Buridan (1985) Theorem IV-13.
[36] Buridan (1985) Theorem IV-16.
[37] Buridan (1985) Theorem IV-17.
[38] Buridan (1985) Theorem IV-18.

Hughes notes some differences between the *Tractatus de consequentiis* and the *Summulae*.[39] (I mention only those moods that use Definitions 9.1 to 9.10.)

(1) The *Summulae* accepts first- and second-figure XLM moods, which are not mentioned in the *Consequentiis*.[40] In a context that included the standard laws of modal subordination, we would have expected these moods to be valid as a consequence of the first- and second-figure non-modal syllogisms. In Buridan's semantics, modal subordination holds only from universal negative necessities and only to particular affirmative possibilities. However, no syllogism has both a universal negative minor and a particular affirmative conclusion; so modal subordination does deliver any of the eight moods covered by Theorems 9.9 and 9.9a.

Theorem 9.9. The XLM moods are valid in Figure 1. Barbara XLM is valid: if $b \rightarrow a$ and $c^{\dagger} \rightarrow b^{*}$ then by interpolating Axioms 1.7 and 1.8 we may form the chain '$c^{\dagger} \rightarrow b^{*} \rightarrow b \rightarrow a \rightarrow a^{\dagger}$', which simplifies to '$c^{\dagger} \rightarrow a^{\dagger}$'. Celarent XLM is also valid: if $b \mid a$ and $c^{\dagger} \rightarrow b^{*}$ then by interpolating Axiom 1.7 we may form the chain '$c^{\dagger} \rightarrow b^{*} \rightarrow b \mid a \leftarrow a^{*}$', which simplifies to '$c^{\dagger} \mid a^{*}$'. Darii XLM and Ferio XLM are similarly valid.

Theorem 9.9a. The XLM moods are valid in Figure 2.

Theorem 9.9b. The LLX moods are valid in Figure 3. It should be mentioned here that Aristotle appears not to adopt a systematic approach to syllogisms with strengthened premises or weakened conclusions; but I see no reason to doubt his full acceptance of modal subordination.[41] In that case he is committed to accepting all the syllogisms mentioned in Theorems 9.9 to 9.11.

(2) The *Consequentiis* accepts Camestres and Baroco XLX (moods that are not mentioned in the *Summulae*).[42] The validity of Camestres XLX follows from that of non-modal Camestres by virtue of modal subordination. Baroco is also valid: if $b \rightarrow a$ and $c^{\dagger} \smile d \mid a^{\dagger}$ then by interpolating Axiom 1.8 we may form the chain '$c \rightarrow c^{\dagger} \smile d \mid a^{\dagger} \leftarrow a \leftarrow b$', which simplifies to '$c \smile d \mid b$'. The remaining second-figure XLX moods are also valid. Cesare: if $b \mid a$ and $c^{\dagger} \rightarrow a^{*}$ then by interpolating Axioms 1.7 and 1.8 we may form the chain '$c \rightarrow c^{\dagger} \rightarrow a^{*} \rightarrow a \mid b$', which simplifies to '$c \mid b$'. Similarly for Festino. Thus we have the general result:

---

[39] Hughes (1989) p.111.
[40] Buridan (2001) 5.7.3 rules 2, 3, 4; p.355.
[41] Thom (1996) pp.34-35.
[42] Buridan (1985) Theorem IV–17; 4.2.59.

Theorem 9.10. The XLX moods are valid in Figure 2. Aristotle explicitly accepts Cesare XLX.[43]

We infer two corollaries by indirect reduction.

[**Theorem 9.10a.** The XXM moods are valid in Figure 1.]

[**Theorem 9.10b.** The LXX moods are valid in Figure 3.]

(3) The *Summulae* accepts Disamis XLX[44] (a mood that is not mentioned in the *Consequentiis*). This mood would be valid if modal subordination was valid for universal affirmative necessities (which it is not); nonetheless the mood is valid.

Theorem 9.11. Disamis XLX is valid. If $c \smile a$ and $c^{\dagger} \rightarrow b^{*}$ then by interpolating Axioms 1.7 and 1.8 we may form the chain '$b \leftarrow b^{*} \leftarrow c^{\dagger} \leftarrow c \smile a$', which simplifies to '$b \smile a$'. Datisi and Ferison XLX are not valid because there is no genuine middle term. By contrast, Aristotle explicitly accepts Felapton XLX,[45] and his comments apply equally to Ferison and Datisi XLX. Bocardo is invalid: supposing that no-one is writing, but humans exist, it is true that not every writer is human, and that every possible writer is necessarily human, but it is also true that every human is human.

[**Theorem 9.11a.** Celarent XLX is valid.]

[**Theorem 9.11b.** Festino XXM is valid.]

(4) The *Summulae* rejects, but the *Consequentiis* accepts, Celarent XLX (Theorem 9.11a).[46] The *Summulae* proposes a counter-example:

> For let us assume that no moon is lacking light and that every heavenly body lacking light is necessarily the moon; still, the conclusion that says that a heavenly body lacking light is not lacking light would be false.

This counter-example fails, as Klima points out.[47] He points out that if the premises are true then the conclusion's subject is null and thus the conclusion is true. The *Consequentiis* offers a proof of the mood's validity; and in fact it is valid. If $b \mid a$ and $c^{\dagger} \rightarrow b^{*}$ then we may form the chain '$c \rightarrow c^{\dagger} \rightarrow b^{*} \rightarrow b \mid a$', which simplifies to '$c \mid a$'. Are the remaining first-figure XLX moods valid? Barbara XLX is not valid, because the minor term in its premise is ampliated to the possible, whereas the conclusion requires the minor term to be unampliated and non-null; so counter-examples can be constructed in which the minor premise is true even though the minor term

---

43 Aristotle (1989) A10, 30b19-20.

44 Buridan (2001) 5.7.3 rule 6; pp.358-359.

45 Aristotle (1989) A11, 31a40-b2.

46 Buridan (2001) 5.7.3 rule 2; p.355. Buridan (1985) Theorem IV-15; 4.2.51.

47 Buridan (2001) p.356 note 44.

is null. Similarly, Darii and Ferio XLX are invalid because the premises imply only that some possible *c* – not that some actual *c* – is or is not *a*.

As can be seen from these examples, we cannot choose one of Buridan's texts and claim unequivocally that its treatment of modal syllogistic is superior to that of the other. What I shall do is to take the best from both treatises, as summarized in Theorems 9.6 to 9.8 and 9.9 to 9.11.

Table 9.2 Buridan's system: L/X/M combinations[48]

Figure	*Combinations*
1	XLM, XXM, Darii Ferio LXL, Celarent LMX, Celarent XLX, Darii Ferio MXM
2	XLM, XLX, Festino LXL, Cesare Camestres LMX, Festino XXM, Cesare Camestres MLX
3	LLX, LXX, Datisi Ferison LXL, Disamis XLX, Datisi XXM, Datisi Ferison MXM

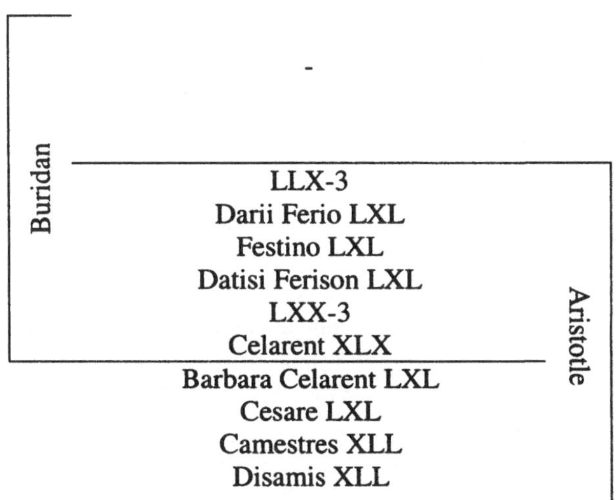

Figure 9.3 Buridan's L/X/M system and Aristotle's system

[48] See Appendix V in Lagerlund (2000) p.239, Table 32.3 in Thom (1996) p.241, and Table 4a in Thom (1997) p.696 – which however is wrongly labelled 'Ockham'.

Buridan considers the effect of taking assertoric propositions to be *simpliciter*, i.e. necessary (conditional on the non-emptiness of the subject-term).[49] This produces Theorems 9.12 to 9.17.

Theorem 9.12. The LXL moods with *simpliciter* assertoric are valid in Figure 1.[50] Barbara: if $b^\dagger \to a^*$ and $\Box\, c \to b$ then we may replace the minor by '$c^\dagger \to b^\dagger$' (Axioms 1.14 and 1.11), and form the chain '$c^\dagger \to b^\dagger \to a^*$', which simplifies to '$c^\dagger \to a^*$'. Celarent: if $b^\dagger \mid a^\dagger$ and $\Box\, c \to b$ then we may replace the minor by '$c^\dagger \to b^\dagger$' and form the chain '$c^\dagger \to b^\dagger \mid a^\dagger$' which simplifies to '$c^\dagger \mid a^\dagger$'. The validity of Darii and Ferio with *simpliciter* minor can be proved in similar ways.

Theorem 9.12a. The MXM moods with *simpliciter* assertoric are valid in Figure 3.[51]

Theorem 9.13. The XLL moods with *simpliciter* assertoric are valid in Figure 1.[52] Barbara: if $\Box\, b \to a$ and $c^\dagger \to b^*$ then we may replace the major by '$b^* \to a^*$' (Axioms 1.12 and 1.11) and form the chain '$c^\dagger \to b^* \to a^*$', which simplifies to '$c^\dagger \to a^*$'. Celarent: if $\Box\, b \mid a$ and $c^\dagger \to b^*$ then we may replace the major by '$b^* \mid a^\dagger$' (Axioms 1.13 and 1.11) and form the chain '$c^\dagger \to b^* \mid a^\dagger$', which simplifies to '$c^\dagger \mid a^\dagger$'. Darii and Ferio can be proved in similar fashion. Barbara XLL does not hold for Avicenna. That would require that if $b^\dagger \to a^\dagger$ and $c^\dagger \to b^*$ then $c^\dagger \to a^*$, but all that these premises imply is that $c^\dagger \to a^\dagger$. This brings to light an important difference between Avicenna's reading of assertorics and the *simpliciter* reading: a universal affirmative *simpliciter* assertoric implies not only that $b^\dagger \to a^\dagger$ but also that $b^* \to a^*$.

[**Theorem 9.13a.** The XMM moods with *simpliciter* assertoric are valid in Figure 2.]

Theorem 9.14. The XMM moods with *simpliciter* assertoric are valid in Figure 1.[53] Barbara: if $\Box\, b \to a$ and $c^\dagger \to b^\dagger$ then we may replace the major by '$b^\dagger \to a^\dagger$' (Axioms 1.14 and 1.11) and form the chain '$c^\dagger \to b^\dagger \to a^\dagger$', which simplifies to '$c^\dagger \to a^\dagger$'. Buridan clearly has a proof like this in mind when he observes:

> But there is no counterinstance wherein the major is about actuality absolutely <*de inesse simpliciter*>, the reason being that a proposition about actuality that

---

49 See Lagerlund (2000) pp.153-154. Buridan's concept of the *simpliciter* is wider than Ockham's, including all of what Ockham calls 'superior/inferior' predications (Definition 8.19). He reverts, in effect, to Kilwardby's concept of the *simpliciter* (Definition 6.9).

50 Buridan (1985) Theorem IV-19.

51 Buridan (1985) Theorem IV-14. Buridan states the result only for Disamis and Bocardo.

52 *Ibid.*

53 Buridan (1985) Theorem IV-11.

is necessary entails a proposition about possibility. For example, the following [inference is valid]: 'This is necessary: "Every B is A"; therefore, every B can be A', and then we would have a perfect syllogism concluding to one universal about possibility from premises that are both about possibility.[54]

Celarent: if $\Box\, b \mid a$ and $c^{\dagger} \rightarrow b^{\dagger}$ then we may replace the major by '$b^{\dagger} \mid a^{*}$' (Axioms 1.15 and 1.11) and form the chain '$c^{\dagger} \rightarrow b^{\dagger} \mid a^{*}$' which simplifies to '$c^{\dagger} \mid a^{*}$'. Darii and Ferio can be proved in a similar fashion.
Theorem 9.14a. The XLL moods with *simpliciter* assertoric are valid in Figure 2.[55]
Theorem 9.15. The LXL moods with *simpliciter* assertoric are valid in Figure 2.[56] Cesare: if $b^{\dagger} \mid a^{\dagger}$ and $\Box\, c \rightarrow a$ then we may replace the minor by '$c^{\dagger} \rightarrow a^{\dagger}$' (Axioms 1.14 and 1.11) and form the chain '$c^{\dagger} \rightarrow a^{\dagger} \mid b^{\dagger}$' which simplifies to '$c^{\dagger} \mid b^{\dagger}$'. Festino similarly. Camestres: if $b^{\dagger} \rightarrow a^{*}$ and $\Box\, c \mid a$ then we may replace the minor by '$c^{\dagger} \mid a^{*}$' (Axioms 1.15 and 1.11) and form the chain '$c^{\dagger} \mid a^{*} \leftarrow b^{\dagger}$' which simplifies to '$c^{\dagger} \mid b^{\dagger}$'. Baroco similarly.
Theorem 9.15a. Datisi and Ferison XMM with *simpliciter* assertoric are valid.[57]
Theorem 9.16. The LXL moods with *simpliciter* assertoric are valid in Figure 3.[58] Disamis: if $c^{\dagger} \smile a^{*}$ and $\Box\, c \rightarrow b$ then we may replace the minor by '$c^{\dagger} \rightarrow b^{\dagger}$' (Axioms 1.14 and 1.11) and form the chain '$b^{\dagger} \leftarrow c^{\dagger} \smile a^{*}$' which simplifies to '$b^{\dagger} \smile a^{*}$'. Bocardo: if $c^{\dagger} \smile d \mid a^{\dagger}$ and $\Box\, c \rightarrow b$ then we may replace the minor by $c^{\dagger} \rightarrow b^{\dagger}$ and form the chain '$b^{\dagger} \leftarrow c^{\dagger} \smile d \mid a^{\dagger}$' which simplifies to '$b^{\dagger} \smile d \mid a^{\dagger}$'. Datisi and Ferison similarly.
[Theorem 9.16a. The MXM moods with *simpliciter* assertoric are valid in Figure 1.]

Some moods that were valid on the ordinary reading of assertorics become invalid on the *simpliciter* reading. These include Barbara and Celarent XLX, and the LLX and LXX moods in Figure 3: a *de dicto* necessity follows only from two *de dicto* necessities.

In addition, the *Summulae* accepts Disamis XLL; and we may broaden this result to include Bocardo XLL.

Theorem 9.17. Disamis and Bocardo XLL are valid if the major is *simpliciter*.[59] Disamis: if $\Box\, c \leftarrow d$ and $\Box\, d \rightarrow a$, and $c^{\dagger} \rightarrow b^{*}$ then we may replace the first of these premises by '$d \rightarrow c$' (Axiom 1.11) and the second by '$d^{*} \rightarrow a^{*}$' (Axioms 1.12 and 1.11), and then by interpolating Axioms 1.7

[54] Buridan (2001) 5.7.2; p.354.
[55] Buridan (1985) Theorem IV-11.
[56] Buridan (1985) Theorem IV-19.
[57] Buridan (1985) Theorem IV-14.
[58] Buridan (1985) Theorem IV-20.
[59] Buridan (2001) 5.7.3; p.359.

and 1.8 we may form the chain '$b^{\dagger} \leftarrow b \leftarrow b^{*} \leftarrow c^{\dagger} \leftarrow c \leftarrow d \leftarrow d^{*} \rightarrow a^{*}$' which simplifies to '$b^{\dagger} \leftarrow d \rightarrow a^{*}$'. Bocardo: if $\Box\, c \leftarrow d$ and $\Box\, d \mid a$, and $c^{\dagger} \rightarrow b^{*}$ then we may replace the first of these premises by '$d \rightarrow c$' (Axiom 1.11) and the second by '$d^{*} \mid a^{\dagger}$' (Axioms 1.13 and 1.11), and then by interpolating Axioms 1.7 and 1.8 we may form the chain '$b^{\dagger} \leftarrow b \leftarrow b^{*} \leftarrow c^{\dagger} \leftarrow c \leftarrow d \leftarrow d^{*} \mid a^{\dagger}$' which simplifies to '$b^{\dagger} \leftarrow d \mid a^{\dagger}$'.

Table 9.3 Buridan's *simpliciter* system: L/X/M combinations

Figure	*Combinations*
1	LXL, XLL, XMM, MXM
2	LXL, XLL, XMM
3	LXL, Disamis Bocardo XLL, Datisi Ferison XMM, Disamis Bocardo MXM

Figure 9.4 Buridan's *simpliciter* L/X/M system and Aristotle's system

Contingency-syllogistic

Buridan outlines a syllogistic for contingency-propositions. He states the following results.

Theorem 9.18. The QMQ moods are valid in Figure 1.[60] Barbara (Celarent): if $b^{\dagger} \rightarrow a^{\ddagger}$ and $c^{\dagger} \rightarrow b^{\ddagger}$ then by interpolating Axiom 1.9 we may

[60] Buridan (1985) Theorem IV-23. King translates '... with a contingent major ... there follows a similar conclusion ... which is *de possibili*, if [the minor] is

form the chain '$c^{\dagger} \rightarrow b^{\ddagger} \rightarrow b^{\dagger} \rightarrow a^{\ddagger}$' which simplifies to '$c^{\dagger} \rightarrow a^{\ddagger}$'. Darii (Ferio): if $b^{\dagger} \rightarrow a^{\ddagger}$ and $c^{\dagger} \smile b^{\ddagger}$ then, interpolating Axiom 1.9, we may form the chain '$c^{\dagger} \smile b^{\ddagger} \rightarrow b^{\dagger} \rightarrow a^{\ddagger}$' which simplifies to '$c^{\dagger} \smile a^{\ddagger}$'.

Theorem 9.19. The QMQ moods are valid in Figure 3.[61] Datisi (Ferison): if $c^{\dagger} \rightarrow a^{\ddagger}$ and $c^{\dagger} \smile b^{\dagger}$ then we have the chain '$b^{\dagger} \smile c^{\dagger} \rightarrow a^{\ddagger}$' which simplifies to '$b^{\dagger} \smile a^{\ddagger}$'. Disamis (Bocardo): if $c^{\dagger} \smile a^{\ddagger}$ and $c^{\dagger} \rightarrow b^{\dagger}$ then we have the chain '$b^{\dagger} \leftarrow c^{\dagger} \smile a^{\ddagger}$' which simplifies to '$b^{\dagger} \smile a^{\ddagger}$.

He states that an M-premise may be replaced by the corresponding Q-premise, and a Q-conclusion by the corresponding M-conclusion.[62] Consequently, we have the following results.

Theorem 9.20. The QQQ moods are valid in Figure 1.[63]

Theorem 9.21. The QQQ moods are valid in Figure 3.[64]

Table 9.4 Buridan's system: Q/Q combinations

Figure	*Combinations*
1	QQQ
2	-
3	QQQ

These results accord with Aristotle's, but it is noteworthy that Buridan secures the validity of the uniform contingency-moods without having recourse to Aristotle's stratagem of ampliating to the contingent.

Let us turn to the X/Q combinations.

Theorem 9.22. Darii (Ferio) QXQ is valid.[65] If $b^{\dagger} \rightarrow a^{\ddagger}$ and $c \smile b$, then, interpolating Axiom 1.8, we can construct the chain '$c^{\dagger} \leftarrow c \smile b \rightarrow b^{\dagger} \rightarrow a^{\ddagger}$', which simplifies to '$c^{\dagger} \smile a^{\ddagger}$. Barbara (Celarent) QXQ is not valid because the premises imply only that $c \rightarrow a^{\ddagger}$ not that $c^{\dagger} \rightarrow a^{\ddagger}$.

Theorem 9.23. Datisi (Ferison) QXQ is valid.[66] If $c^{\dagger} \rightarrow a^{\ddagger}$ and $c \smile b$ then, interpolating Axiom 1.8, we may form the chain '$b^{\dagger} \leftarrow b \smile c \rightarrow c^{\dagger} \rightarrow a^{\ddagger}$'

---

contingent'. But surely this should be the other way round: there follows a similar conclusion ... which is contingent, if the minor is *de possibili*. It would be consistent with Buridan's systematic approach to establish QMQ moods first, only later deriving QQQ moods by strengthening the minor premise.

61 *Ibid.*

62 Buridan (1985) Theorems IV-21 and IV-22.

63 Buridan (2001) 5.6.6.

64 Buridan (2001) 5.6.6.

65 Buridan (1985) Theorem IV-24.

66 Buridan (1985) Theorem IV-25.

which simplifies to '$b^\dagger \smile a^\ddagger$. The other Figure 3 QXQ moods are invalid: e.g., if for some 'd', $c^\dagger \leftarrow d^\dagger \rightarrow a^\ddagger$ and $c \rightarrow b$ then no middle term can be constructed.

Table 9.5 Buridan's system: X/Q combinations

Figure	*Combinations*
1	Darii (Ferio) QXQ
2	-
3	Datisi (Ferison) QXQ

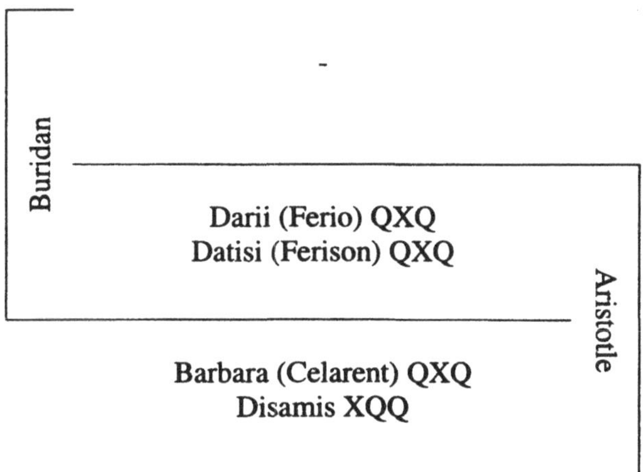

Figure 9.5 Buridan's X/Q system and Aristotle's system

With *simpliciter* assertorics, the results are slightly different.

Theorem 9.24. The QXQ moods with *simpliciter* assertoric are valid in Figure 1. Barbara (Celarent): if $b^\dagger \rightarrow a^\ddagger$ and $\Box\, c \rightarrow b$ then we may replace the minor by '$c^\dagger \rightarrow b^\dagger$' (Axioms 1.14 and 1.11) forming the chain '$c^\dagger \rightarrow b^\dagger \rightarrow a^\ddagger$' which simplifies to '$c^\dagger \rightarrow a^\ddagger$'. Darii (Ferio) can be validated in a similar way.

Theorem 9.25. The QXQ moods with *simpliciter* assertoric are valid in Figure 3. Disamis (Bocardo): if $c^\dagger \smile a^\ddagger$, and $\Box\, c \rightarrow b$ then we may replace the minor by '$c^\dagger \rightarrow b^\dagger$', forming the chain '$b^\dagger \leftarrow c^\dagger \smile a^\ddagger$', which simplifies to '$b^\dagger \smile a^\ddagger$'. Datisi (Ferison): if $c^\dagger \rightarrow a^\ddagger$ and $\Box\, c \leftarrow d$ and $\Box\, d \rightarrow b$ then we may replace the last two conjuncts by '$c^\dagger \leftarrow d^\dagger$' and '$d^\dagger \rightarrow b^\dagger$', forming the chain '$b^\dagger \leftarrow d^\dagger \rightarrow c^\dagger \rightarrow a^\ddagger$', which simplifies to '$b^\dagger \smile d^\dagger \rightarrow a^\ddagger$'. Disamis XQQ is invalid: if for some 'd', $\Box\, c \leftarrow d$ and $\Box\, d \rightarrow a$ (so that $c^\dagger \leftarrow d^\dagger \rightarrow$

$a^\dagger$), and $c^\dagger \rightarrow b^\ddagger$, then we have the chain '$b^\dagger \leftarrow c^\dagger \leftarrow d^\dagger \rightarrow a^\dagger$', simplifying to '$b^\dagger \leftarrow d^\dagger \rightarrow a^\dagger$' but not '$b^\dagger \leftarrow d^\dagger \rightarrow a^\ddagger$'.

Table 9.6 Buridan's *simpliciter* system: X/Q combinations

Figure	*Combinations*
1	QXQ
2	-
3	QXQ

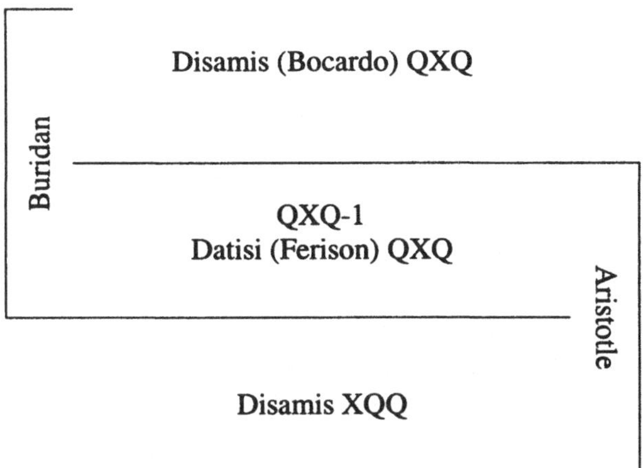

Figure 9.6 Buridan's *simpliciter* X/Q system and Aristotle's system

We now pass to the L/Q combinations.

Theorem 9.26. The QLQ moods in Figures 1 are valid.[67] Barbara (Celarent): if $b^\dagger \rightarrow a^\ddagger$ and $c^\dagger \rightarrow b^*$ then, interpolating Axioms 1.7 and 1.8 we may form the chain '$c^\dagger \rightarrow b^* \rightarrow b \rightarrow b^\dagger \rightarrow a^\ddagger$', simplifying to '$c^\dagger \rightarrow a^\ddagger$'. Darii (Ferio): if $b^\dagger \rightarrow a^\ddagger$ and $c^\dagger \smile b^*$, then we may form the chain '$c^\dagger \smile b^* \rightarrow b \rightarrow b^\dagger \rightarrow a^\ddagger$', simplifying to '$c^\dagger \smile a^\ddagger$'.

Theorem 9.27. The QLQ moods in Figure 3 are valid.[68] Datisi (Ferison): if $c^\dagger \rightarrow a^\ddagger$ and $c^\dagger \smile b^*$ then by interpolation we may form the chain '$b^\dagger \leftarrow b \leftarrow b^* \smile c^\dagger \rightarrow a^\ddagger$' which simplifies to '$b^\dagger \smile a^\ddagger$'. Also valid is the unAristotelian mood Disamis (Bocardo) QLQ: if $c^\dagger \smile a^\ddagger$ and $c^\dagger \rightarrow b^*$, then by interpolation

[67] Buridan (1985) Theorem IV-23.

[68] *Ibid.*

we may form the chain '$b^\dagger \leftarrow b \leftarrow b^* \leftarrow c^\dagger \smile a^\ddagger$', which simplifies to '$b^\dagger \smile a^\ddagger$'.

Theorem 9.28. The LQL moods in Figure 1 are valid.[69] Barbara: if $b^\dagger \rightarrow a^*$ and $c^\dagger \rightarrow b^\ddagger$ then by interpolation we may form the chain '$c^\dagger \rightarrow b^\ddagger \rightarrow b^\dagger \rightarrow a^*$', which simplifies to '$c^\dagger \rightarrow a^*$'. Celarent: if $b^\dagger \mid a^\dagger$ and $c^\dagger \rightarrow b^\ddagger$ then by interpolation we may form the chain '$c^\dagger \rightarrow b^\ddagger \rightarrow b^\dagger \mid a^\dagger$', which simplifies to '$c^\dagger \mid a^\dagger$'. (Alexander had derived the validity of this mood from that of Disamis MMM.[70]) Darii: if $b^\dagger \rightarrow a^*$ and $c^\dagger \smile b^\ddagger$ then by interpolation we may form the chain '$c^\dagger \smile b^\ddagger \rightarrow b^\dagger \rightarrow a^*$', which simplifies to '$c^\dagger \smile a^*$'. Ferio: if $b^\dagger \mid a^\dagger$ and $c^\dagger \smile b^\ddagger$ then by interpolation we may form the chain '$c^\dagger \smile b^\ddagger \rightarrow b^\dagger \mid a^\dagger$', which simplifies to '$c^\dagger \smile b^\dagger \mid a^\dagger$'.

Theorem 9.29. The LQL moods in Figure 2 are valid.[71] Cesare: if $b^\dagger \mid a^\dagger$ and $c^\dagger \rightarrow a^\ddagger$ then by interpolation we may form the chain '$c^\dagger \rightarrow a^\ddagger \rightarrow a^\dagger \mid b^\dagger$', which simplifies to '$c^\dagger \mid b^\dagger$. Camestres: if $b^\dagger \rightarrow a^*$ and $c^\dagger \mid a^\dagger$ then by interpolation we may form the chain '$c^\dagger \mid a^\dagger \leftarrow a \leftarrow a^* \leftarrow b^\dagger$', which simplifies to '$c^\dagger \mid b^\dagger$'. Festino: if $b^\dagger \mid a^\dagger$ and $c^\dagger \smile a^\ddagger$ then by interpolation we may form the chain '$c^\dagger \smile a^\ddagger \rightarrow a^\dagger \mid b^\dagger$', which simplifies to '$c^\dagger \smile a^\dagger \mid b^\dagger$'. Baroco: if $b^\dagger \rightarrow a^*$ and $c^\dagger \smile a^\ddagger$ then by interpolation we may form the chain '$c^\dagger \leftarrow d \rightarrow a^\ddagger \rightarrow a^\dagger \rightarrow b^*$', which simplifies to '$c^\dagger \smile b^*$'.

Theorem 9.30. The LQL moods in Figure 3 are valid.[72] Datisi: if $c^\dagger \rightarrow a^*$ and $c^\dagger \smile b^\ddagger$, then by interpolation we may form the chain '$b^\dagger \leftarrow b^\ddagger \smile c^\dagger \rightarrow a^*$', which simplifies to '$b^\dagger \smile a^*$'. Disamis: if $c^\dagger \smile a^*$, and $c^\dagger \rightarrow b^\ddagger$, then by interpolation we may form the chain '$b^\dagger \leftarrow b^\ddagger \leftarrow c^\dagger \smile a^*$', which simplifies to '$b^\dagger \smile a^*$'. Ferison: if $c^\dagger \mid a^\dagger$ and $c^\dagger \smile b^\ddagger$, then by interpolation we may form the chain '$b^\dagger \leftarrow b^\ddagger \smile c^\dagger \mid a^*$', which simplifies to '$b^\dagger \smile c^\dagger \mid a^*$'. Bocardo: if for some 'd', $c^\dagger \leftarrow d \mid a^\dagger$, and $c^\dagger \rightarrow b^\ddagger$, then by interpolation we may form the chain '$b^\dagger \leftarrow b^\ddagger \leftarrow c^\dagger \leftarrow d \mid a^\dagger$', which simplifies to '$b^\dagger \leftarrow d \mid a^\dagger$'.

Table 9.7 Buridan's system: L/Q combinations

Figure	*Combinations*
1	LQL, QLQ
2	LQL
3	LQL, QLQ

---

[69] Buridan (2001) 5.7.7.

[70] See above pp.31-32.

[71] *Ibid.*

[72] *Ibid.*

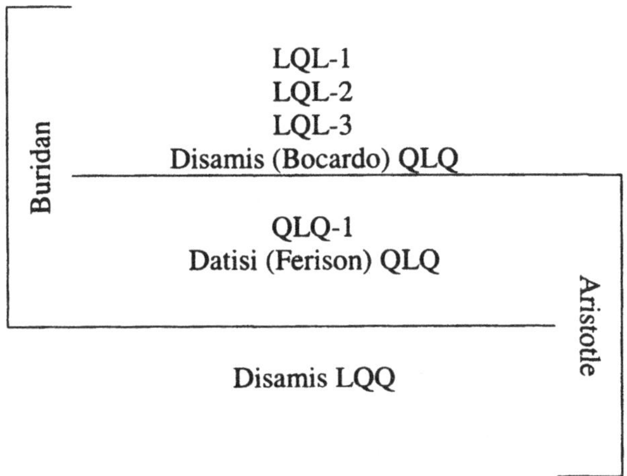

Figure 9.7 Buridan's L/Q system and Aristotle's system

Buridan states no results about L/Q/M or L/X/Q or X/Q/M combinations; but we can draw some inferences from his already stated results, given the relation between Q- and M-propositions.

[**Theorem 9.31.** Celarent LQX is valid.] This follows from the validity of Celarent LMX (Theorem 9.7a.). The Aristotelian Ferio LQX is, however, invalid: if $b^\dagger \mid a^\dagger$ and $c^\dagger \smile b^*$, then the chain '$c^\dagger \smile b^* \to b \to b^\dagger \mid a^\dagger$' delivers the result that $c^\dagger \smile b^\dagger \mid a^\dagger$ but not that $c \smile b^\dagger \mid a$.

[**Theorem 9.32.** Cesare and Camestres LQX are valid.] This follows from the validity of Cesare and Camestres LMX (Theorem 9.6a.) The Aristotelian Festino and Baroco LQX are invalid, because the premises imply only that some possible *c* (not: some actual *c*) is not *b*.

[**Theorem 9.33.** Cesare and Camestres QLX are valid.] This follows from Theorem 9.8b. The Aristotelian Baroco QLX is invalid, because the premises imply only that some possible *c* is not *b*. The Aristotelian Ferison LQX is also invalid, because the premises imply only that some possible *b* is not *a*. Thus, we have the following results.

Table 9.8 Buridan's system: L/X/Q combinations

Figure	*Combinations*
1	Celarent LQX
2	Cesare Camestres LQX, Cesare Camestres QLX
3	-

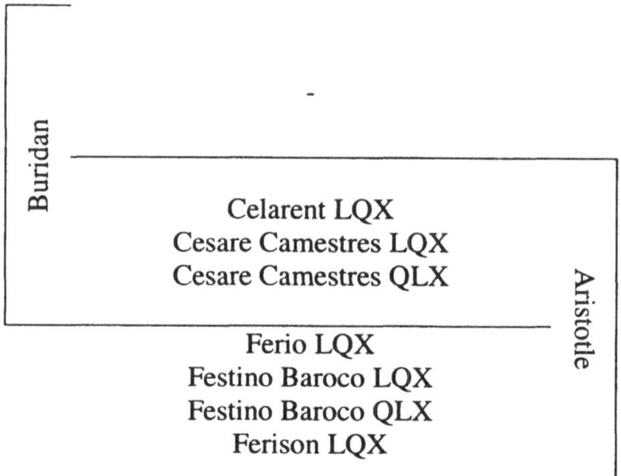

Figure 9.8 Buridan's L/X/Q system and Aristotle's system

From Table 9.1, given that Q-propositions always imply their M-counterparts, we can infer that Buridan is committed to accepting all QLM and LQM moods in all figures; this is Aristotelian doctrine (Table 1.9).

[**Theorem 9.34.** The LQM moods are valid in Figure 1.]

[**Theorem 9.35.** The LQM moods are valid in Figure 2.]

[**Theorem 9.36.** The LQM moods are valid in Figure 3.]

[**Theorem 9.37.** The QLM moods are valid in Figure 1.]

[**Theorem 9.38.** The QLM moods are valid in Figure 2.]

[**Theorem 9.39.** The QLM moods are valid in Figure 3.]

From Table 9.3, given that Q-propositions imply their M-counterparts, we can infer that Buridan is committed to accepting all XQM and QXM moods in Figure 1, all XQM moods in Figure 2, Disamis and Ferison XQM, Disamis and Bocardo QXM – provided that assertoric premises are *simpliciter*. As Buridan puts it, 'If one of the premises is about actuality and the other is about contingency, then we should posit a number of rules analogous to the ones that were given about the mixture of [premises] about actuality and possibility'.[73]

Theorem 9.40. The XQM moods with *simpliciter* assertoric are valid in Figure 1.

Theorem 9.41. The QXM moods with *simpliciter* assertoric are valid in Figure 1.

[73] Buridan (2001) 5.7.4.

Theorem 9.42. The XQM moods with *simpliciter* assertoric are valid in Figure 2.
Theorem 9.43. Datisi and Ferison XQM with *simpliciter* assertoric are valid.
Theorem 9.44. Disamis and Bocardo QXM with *simpliciter* assertoric are valid.

Table 9.9 Buridan's *simpliciter* system: X/Q/M combinations

Figure	*Combinations*
1	XQM, QXM
2	XQM
3	Datisi Ferison XQM, Disamis Bocardo QXM

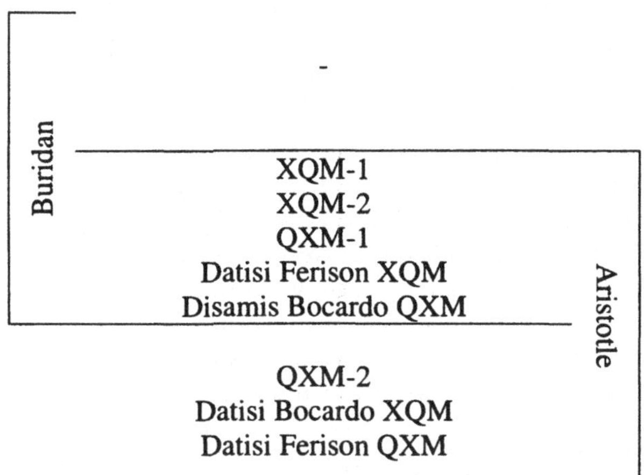

Figure 9.9 Buridan's *simpliciter* X/Q/M system and Aristotle's system

Alternate systems

Like Ockham, Buridan explores a number of alternatives to his principal system. These include systems in which the ampliation of modals is cancelled by prefixing the words 'that is' to the subject,[74] or modals are read in the compound sense.[75]

[74] Buridan (1985) Theorems IV-7 to IV-9.
[75] Buridan (1985) Theorems IV-1 to IV-3.

Restricted modals

When the expression 'what is' is prefixed to the subject of a divisive modal, we have the actualist reading of Abélard and Ockham. Buridan sketches the logic of these propositions, and we shall briefly consider cases where all modal propositions in a syllogism are restricted by the 'what is' phrase.

Theorem 9.45. The LLL moods are valid in Figure 1.[76]
Theorem 9.45a. The LMM moods are valid in Figure 2.[77]
Theorem 9.45b. The MLM moods are valid in Figure 3.[78]
Theorem 9.46. The LXL moods are valid in Figure 1.[79]
Theorem 9.46a. The LMX moods are valid in Figure 2.[80]
Theorem 9.46b. The MXM moods are valid in Figure 3.
Theorem 9.47. The MLM moods are valid in Figure 1.[81]
Theorem 9.47a. The MLM moods are valid in Figure 2.[82]
Theorem 9.47b. The LLL moods are valid in Figure 3.[83]
Theorem 9.48. The MXM moods are valid in Figure 1.[84]
Theorem 9.48a. The MLX moods are valid in Figure 2.[85]
[**Theorem 9.48b.** The LXL moods are valid in Figure 3.]
Theorem 9.49. The LML moods are valid in Figure 3.[86]
[**Theorem 9.49a.** The MMM moods are valid in Figure 1.]
[**Theorem 9.49b.** The MLL moods are valid in Figure 2.]

Ockham is in full agreement with these results.[87]

In addition to these results, the *Summulae* adds a few more.

Theorem 9.50. The QXQ moods are valid in Figure 1.[88]

---

[76] Buridan (1985) Theorem IV-7.
[77] *Ibid.*
[78] Buridan (1985) Theorem IV-9.
[79] Buridan (2001) 5.7.3. rule 1.
[80] Buridan (1985) Theorem IV-8.
[81] Buridan (1985) Theorem IV-7.
[82] *Ibid.*
[83] Buridan (1985) Theorem IV-9.
[84] Buridan (2001) 5.7.2. rule 4.
[85] Buridan (1985) Theorem IV-8.
[86] Buridan (1985) Theorem IV-9.
[87] Ockham (1962) IIIa ch.20:60ff, ch.25:26-30.
[88] Buridan (2001) 5.7.4 rule 4. Klima's translation – 'no universal conclusion can be inferred with the addition of the phrase "that is"' – is incorrect. The Latin in Buridan (1499) reads 'non potest inferri conclusion universalis nisi cum illa additione quod est'.

These results are tabulated in Table 9.10, which should be compared with Ockham's results.

Table 9.10 Buridan's system of restricted modals: L/X/Q/M combinations

Figure	*Combinations*
1	LLL, MLM, MXM, LXL, QXQ
2	LMX, MLX, LMM MLM
3	LLL, MLM, LML

Compound modals

So far as compound modals go, Buridan (like Ockham) accepts all LLL syllogisms in the compound sense but no MMM or QQQ syllogisms.[89]

Summary

Buridan's system is exceptionally elegant, and is the least metaphysical of all the systems we have studied. Buridan's main system is based on a uniform ampliationist understanding of modal propositions and does not require ampliation to the contingent in the Q/Q layer. Like all the logicians we have discussed, Buridan makes use of the basic essentialist assumption that some individuals possess some of their properties essentially, and therefore necessarily. However, he does not in his modal syllogistic use the stronger essentialist assumption that there are Kinds, i.e. that some properties are such that it is necessarily the case that whatever possesses them possesses them essentially.

The system comes in two versions – an ordinary version, and one in which assertorics are read *simpliciter*. The ordinary version provides a poor match with Aristotle's system. The *simpliciter* version can be based on Barbara, Darii, Baroco, Datisi, Disamis, Ferison, Bocardo LLL, the MMM moods in all figures, the LXL and XLL moods in Figures 1 and 2, the LXL moods in Figure 3, the XLX moods in Figure 2, the LLX moods in Figure 3, the QMQ moods in Figures 1 and 3, Barbara, Celarent, Disamis, Bocardo QXQ.

[89] Buridan (1985) Theorem IV-1. See Lagerlund (2000) pp.145-149.

Chapter 10

Conclusion

On the basis of the interpretations offered in the preceding chapters, we can now view the inter-relations holding among the principal medieval modal systems, layer by layer; and we can judge the success of the various systems as interpretations of Aristotle's modal syllogistic.

The systems

L/M systems

There are four systems. Abélard and Ockham accept the LLL moods in Figures 1 and 3. Campsall adds to these Barbara and Darii, Datisi and Disamis MMM. Averröes and Kilwardby follow Aristotle in accepting the LLL moods in all figures. Avicenna and Buridan accept the LLL moods in all figures, plus the MMM moods in Figures 1 and 3. These results are displayed in Table 10.1.

Table 10.1 L/M systems

System	*Base*
Abélard, Ockham	LLL-1, LLL-3
Avicenna, Buridan	Abélard + LLL-2, MMM-1, MMM-3
Campsall	Abélard + LLL-2, Barbara Darii Datisi Disamis MMM
Averröes, Kilwardby	Abélard + LLL-2

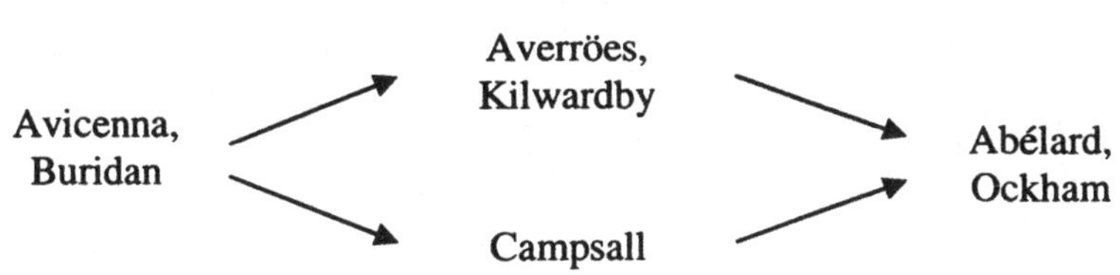

Figure 10.1 Containments among L/M systems

L/X/M systems

Table 10.2 L/X/M systems

System	*Base*
Abélard, Ockham	LXL-1, LXL-3
Avicenna	Abélard + LXL-2, XLL-2
Buridan	LLX-3, Darii Ferio LXL, Festino LXL, Datisi Ferison LXL, LXX-3, Celarent XLX
Buridan *simpliciter*	Avicenna + XLL-1, Disamis Bocardo XLL
Campsall	Celarent Ferio LXL, LXL-2, Ferison Bocardo LXL, Celarent Ferio XLL, XLL-2, Ferison Bocardo XLL
Campsall *simpliciter*, Kilwardby *simpliciter*	Abélard + LXL-2, XLL-1, XLL-2, XLL-3
Ockham *simpliciter*	Abélard + XLL-1, XLL----2

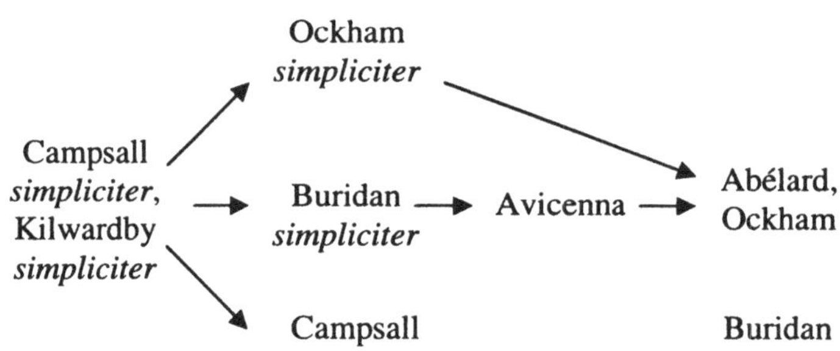

Figure 10.2 Containments among L/X/M systems

Here we find seven distinct systems, falling into three groups. The first group consists solely of Buridan's system, which accepts the particular LXL moods in Figure 1, along with Festino, Datisi and Ferison LXL, the LLX and LXX moods of Figure 3, and Celarent XLX. These LLX, LXX and XLX moods are not provable in other systems without the aid of modal subordination.

In the second group, the weakest system is that of Abélard and Ockham, which accepts the LXL moods in Figures 1 and 3. This base can be expanded in two different ways. (i) Avicenna adds the LXL and XLL moods in Figure 2; and to Avicenna's system, Buridan's *simpliciter* system adds the XLL moods in Figure 1 and Disamis and Bocardo XLL. (ii) Ockham's *simpliciter* system adds to Abélard's base the XLL moods in Figures 1 and 2. The (maximal) *simpliciter* system of Kilwardby and Campsall can be seen as making additions to either Buridan's or Ockham's *simpliciter* system. To Buridan's *simpliciter* system it adds Datisi and Ferison XLL; to Ockham's it adds the LXL moods of Figure 2 and the XLL moods of Figure 3.

The third group has a single member – Campsall's system – which is based on Celarent and Ferio LXL and XLL, the LXL and XLL moods in Figure 2, Ferison and Bocardo LXL, Ferison and Bocardo XLL. See Table 10.2.

Q/Q systems

Buridan, Campsall, Kilwardby and Ockham all agree in accepting the uniform contingency-syllogisms in Figures 1 and 3.

Table 10.3 Q/Q systems

System	*Base*
Buridan, Campsall, Kilwardby, Ockham	QQQ-1, QQQ-3

X/Q systems

There are three systems. Buridan accepts Darii (Ferio) and Datisi (Ferison) QXQ. Buridan and Campsall (using *simpliciter* assertorics), Kilwardby (using *simpliciter* assertorics and natural contingencies), and Ockham, add the remaining first- and third-figure QXQ moods. Ockham's *simpliciter* system (subject to the acceptance of Ockham's Thesis)[1] additionally includes Barbara and Darii XQQ.

[1] Theorem 8.12.

Table 10.4 X/Q systems

System	*Base*
Buridan	Darii (Ferio) Datisi (Ferison) QXQ
Buridan *simpliciter*, Campsall *simpliciter*, Kilwardby natural *simpliciter*, Ockham	Buridan + Barbara (Celarent) QXQ, Disamis (Bocardo) QXQ
Ockham *simpliciter*	Buridan *simpliciter* + Barbara Darii XQQ

Ockham *simpliciter* ⟶ Buridan *simpliciter*, Campsall *simpliciter*, Kilwardby natural *simpliciter*, Ockham ⟶ Buridan

Figure 10.3 Containments among X/Q systems

L/Q systems

There are two systems. Campsall, Kilwardby (using natural contingency) and Ockham accept the first- and third-figure QLQ moods. Buridan additionally accepts the LQL moods in all figures.

Table 10.5 L/Q systems

System	*Base*
Buridan	Campsall + LQL-1 LQL-2 LQL-3
Campsall, Kilwardby natural, Ockham	QLQ-1 QLQ-3

Buridan ⟶ Campsall, Kilwardby natural, Ockham

Figure 10.4 Containments among L/Q systems

L/X/Q systems

There are four systems. Kilwardby's 'natural' system accepts Celarent and Ferio LQX, Cesare and Festino LQX, and Camestres and Baroco QLX. Buridan accepts Celarent, Cesare and Camestres LQX, along with Cesare and Camestres QLX. Ockham accepts all LQX and QLX moods in Figure 2. Campsall accepts all LQX and QLX moods in all figures.

Table 10.6 L/X/Q systems

System	*Base*
Buridan	Celarent Cesare Camestres LQX, Cesare Camestres QLX
Campsall	LQX-1, LQX-2, LQX-3, QLX-1, QLX-2, QLX-3
Kilwardby natural	Celarent Ferio LQX, Cesare Festino LQX, Camestres Baroco QLX
Ockham	LQX-2, QLX-2

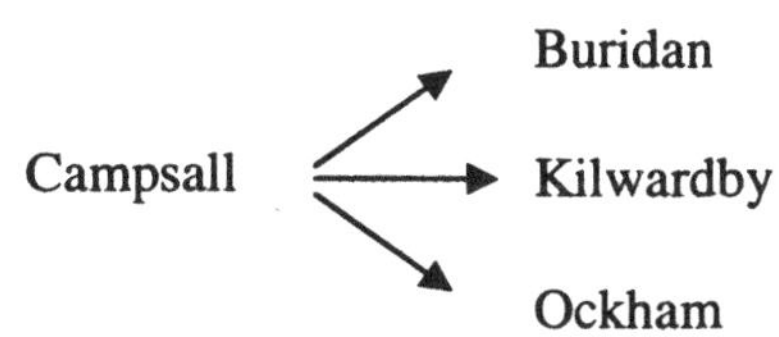

Figure 10.5 Containments among L/X/Q systems

L/Q/M systems

Here all authors, except Ockham, agree in accepting all LQM and QLM moods in all figures. Ockham accepts the QLM moods in all figures, but the LQM moods only in Figure 2.

Table 10.7 L/Q/M systems

System	*Base*
Campsall, Buridan, Kilwardby indefinite	LQM-1, LQM-2, LQM-3, QLM-1, QLM-2, QLM-3
Ockham	LQM-2, QLM-1, QLM-2, QLM-3

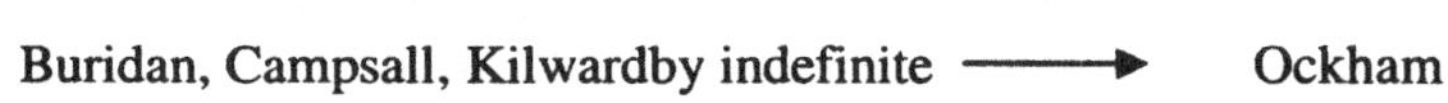

Figure 10.6 Containments among L/Q/M systems

X/Q/M systems

There are four systems. Ockham's ordinary system consists of the QXM moods in Figures 1 and 3. His *simpliciter* system adds the XQM moods in Figure 1 and Cesare and Festino XQM. Buridan's *simpliciter* system accepts the QXM moods in Figure 1 plus Disamis and Bocardo QXM, the XQM moods in Figures 1 and 2, and Datisi and Ferison XQM. Campsall's system and Kilwardby's indefinite *simpliciter* system favour the maximal set comprising all XQM and QXM moods in all figures.

Table 10.8 X/Q/M systems

System	*Base*
Buridan *simpliciter*	XQM-1, XQM-2, Datisi Ferison XQM, QXM-1, Disamis Bocardo QXM
Campsall, Kilwardby indefinite *simpliciter*	XQM-1, XQM-2, XQM-3, QXM-1, QXM-2, QXM-3
Ockham	QXM-1, QXM-3
Ockham *simpliciter*	Ockham + XQM-1, Cesare Festino XQM

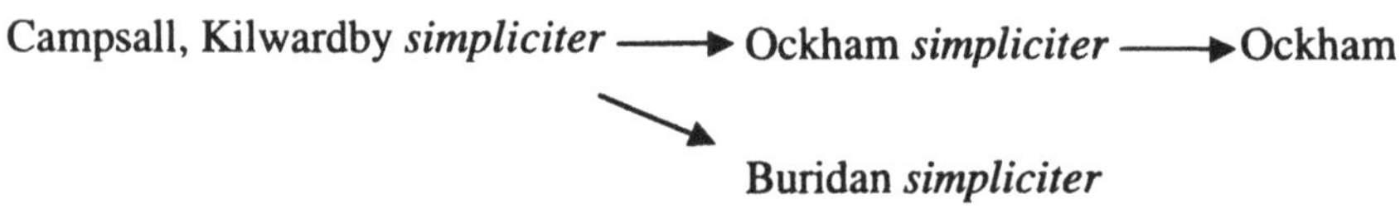

Figure 10.7 Containments among X/Q/M systems

In considering these relations of containment, it is important to bear in mind the different meanings attached to modal and non-modal propositions in the different systems. Some of these are displayed in Table 10.9. As can be seen from that Table, the different systems to a greater or lesser extent employ notions drawn from Aristotelian essentialism. The first of these essentialist notions is that some features attach to some individuals essentially and thus necessarily; in other words that there are *de re* necessities. The second is that some features necessarily belong to their subjects only essentially; in other words that there are Kinds. In principle, three types of system can therefore be distinguished. The first type would be systems that do not assume that there are *de re* necessities (and thus do not assume that there are Kinds). The second type would be systems assuming that there are *de re* necessities but not that there are Kinds. The third type would be systems assuming that there are Kinds (and therefore that there are *de re* necessities). None of the systems we have discussed is of the first type. Type 2 systems include those of Abélard, Avicenna,

Avicenna descriptional, Kilwardby, Kilwardby *simpliciter*, Campsall, Ockham, Ockham superior/inferior, Buridan, and Buridan *simpliciter*. The remaining systems (Averröes, Campsall *simpliciter* or first-class *ut nunc*, and Ockham *simpliciter*) are of type 3. And some of these type 3 systems employ deeper essentialist tenets. For instance, Campsall appears to rely on the postulate (Postulate 7.1) that for every term 'a' there is a *per se* term 'S_a' such that $a \leftrightarrow S_a$; and Ockham at one point seems to endorse the thesis (Theorem 8.12) that if 'a' is *per se* then if $x \rightarrow a^\dagger$ then $x \rightarrow a$.

Table 10.9 Semantics for different systems

System	*La*	*Le*	*Xa*
Abélard	$b \rightarrow a^*$	$b \mid a^\dagger$	$b \rightarrow a$
Averröes	$\Box B \rightarrow A$	$\Box B \mid A$	$\Diamond b \nrightarrow b^*$ and $\Box a \rightarrow a^*$ and $\Box b \rightarrow a$
Avicenna	$b^\dagger \rightarrow a^*$	$b^\dagger \mid a^\dagger$	$b^\dagger \rightarrow a^\dagger$
Avicenna$_D$	$b^\dagger \rightarrow (a/b)^*$	$b^\dagger \mid (ab)^\dagger$	-
Buridan	$b^\dagger \rightarrow a^*$	$b^\dagger \mid a^\dagger$	$b \rightarrow a$
Buridan$_S$	$b^\dagger \rightarrow a^*$	$b^\dagger \mid a^\dagger$	$\Box b \rightarrow a$
Campsall	$\Box B \rightarrow a$	For some 'd', 'e', $b \rightarrow D$ and $\Box d \mid e$ and $E \leftarrow a$	$b \rightarrow a$
Campsall$_S$	$\Box B \rightarrow a$	For some 'd', 'e', $b \rightarrow D$ and $\Box d \mid e$ and $E \leftarrow a$	$\Box B \rightarrow a$
Kilwardby	$\Box b \rightarrow a$	$\Box b^\dagger \mid a^\dagger$	$b \rightarrow a$
Kilwardby$_N$	$\Box b \rightarrow a$	$\Box b^\dagger \mid a^\dagger$	$\Box b^\dagger \rightarrow a^\dagger$
Kilwardby$_S$	$\Box b \rightarrow a$	$\Box b^\dagger \mid a^\dagger$	$\Box b \rightarrow a$
Ockham	$b \rightarrow a^*$	$b \mid a^\dagger$	$b \rightarrow a$
Ockham$_S$	$b \rightarrow a^*$	$b \mid a^\dagger$	$\Box B \rightarrow a$
Ockham$_I$	$b \rightarrow a^*$	$b \mid a^\dagger$	$\Box b \rightarrow a$

Abbreviations: D (descriptional), I (superior/inferior), N (natural), S (*simpliciter*)

The interpretation of Aristotle's modal syllogistic

Some of these systems replicate Aristotle's results. In the L/M layer Averröes and Kilwardby achieve a perfect match with Aristotle. In the Q/Q combinations, Kilwardby, Campsall, Ockham and Buridan are able to replicate Aristotle's results; in the L/Q/M combinations Kilwardby, Campsall, and Buridan; in the X/Q/M layer Kilwardby and Campsall. Elsewhere the medievals diverge from Aristotle to a greater or lesser extent.

An interpretation of a text can be judged successful or unsuccessful against a number of distinct criteria. Among these, three stand out – comprehensiveness, unity and congeniality. Firstly, an interpretation may represent the text comprehensively, or may be selective in its coverage. Secondly, an interpretation may employ uniform means of representation to account for uniform elements in the text, or it may segment the text, using different means of representation for the same elements in different segments. Thirdly, an interpretation's means of representation may be congenial with the material in the text, or may be comparatively alien to it.

In terms of the comprehensiveness with which they fit the data of Aristotle's text, Kilwardby's is the best of the medieval interpretations. His interpretations exactly fit Aristotle's results in four out of the eight layers – the L/M, Q/Q, L/Q/M and X/Q/M layers. Kilwardby's departures from Aristotle are not numerous, except for the L/X/M layer where he accepts fourteen moods rejected by Aristotle. In terms of unity, his interpretations do not fare well: necessity-propositions are read differently in the L/M and the L/X/M moods, contingency-propositions are read in three different ways in the Q/Q, the Q/L/M, and the other contingency-moods; and assertorics are read in two different ways. Arguably, Kilwardby's interpretation evinces a deep congeniality with Aristotle's text by recognizing the notions of the *per se* and species as implicit in that text. At the same time, Kilwardby's interpretive framework seems to me alien to Aristotle's text to the extent that it embraces the notion that assertoric propositions ought to be generally required to be *simpliciter*. I find no support for this notion in the *Analytics*, where the requirement for assertorics to be *simpliciter* is brought in to cover what Aristotle evidently sees as an exceptional case. In non-exceptional cases, his counter-examples to invalid syllogisms often make use of assertoric propositions such as 'Every animal is moving'[2] that could at best be true *ut nunc*.

Campsall achieves a perfect fit with Aristotle's results in three layers (Q/Q, L/Q/M and X/Q/M). But his departures from Aristotelian doctrine are considerable. In the L/X/M layer he accepts 14 moods more than Aristotle, in the L/X/Q layer 15. At the same time, his approach is quite highly unified, using a single understanding of modal propositions in all layers except for the uniform contingency-syllogisms. Its congeniality with

[2] Aristotle (1989) A9, 30a23-25; 30b2-6; A11, 32a4-5.

Aristotle's text is however open to serious question. Aristotle could not have accepted Campsall's Thesis (Theorem 7.1) that what is possible is actual; some Aristotelian possibilities are sometimes unactualized. (Similarly, Avicenna's reduction of assertorics to possibilities does not cohere with Aristotle's thoughts on possibility.)

The interpretations of Ockham and Buridan are not very comprehensive, Buridan getting only two of the Aristotelian layers right (Q/Q and L/M/Q), Ockham only one (Q/Q). In the L/M/X layer Ockham (like Abélard) manages to endorse only six out of Aristotle's ten syllogisms; Buridan only five. But both Ockham's and Buridan's interpretations are highly unified. Like Campsall, Ockham falls short of complete unity by using different readings of contingency-propositions in the Q/Q and other Q layers. Buridan's system is the most unified; he uses a single ampliated understanding of all modal propositions in all contexts. However, Buridan's (and Avicenna's) idea that all modal propositions ought to be ampliated to the possible hardly seems to be congenial with Aristotle's explicit treatment of some contingency-propositions as unampliated.[3]

Does all this betoken a series of failed interpretations? If so, that may be because the task of constructing a philosophically satisfying interpretation of Aristotle's modal syllogistic was always an impossible one. The text to be interpreted is radically incomplete, lacking any semantic base. It also appears to contain a number of inconsistencies. How then could its interpretations ever be other than selective, segmented, or alien? What can be said is that, if the medieval logicians failed in their interpretive task, they succeeded mightily in exercising their logical genius to create some fascinating systems at the intersection of modal theory and essentialist metaphysics.

[3] Aristotle (1989) A13, 32b23-32.

Bibliography

Abélard, P. (1927), *Logica Ingredientibus* in Geyer, B. (ed.), *Abaelards philosophischen Schriften*, I, 1-3, Münster.

Abélard, P. (1956), *Dialectica*, edited by L.M. De Rijk, Van Gorcum, Assen.

Abélard, P. (1958), *Petri Abaelardi Super Perermenias, XII-XIV* = L. Minio-Paluello, (ed.), *Twelfth Century Logic texts and studies II*, Edizioni di storia e letteratura, Roma.

Ackrill, J.L. (1973), *Aristotle's Ethics*, Faber & Faber, London.

Alexander of Aphrodisias (1991), *On Aristotle's **Prior Analytics** 1.1-7*, translated by Jonathan Barnes, Susanne Bobzien, Kevin Flannery S.J. and Katerina Ierodiakonou, Cornell University Press, Ithaca.

Alexander of Aphrodisias (1999a), *On Aristotle's **Prior Analytics** 1.8-13*, translated by Ian Mueller with Josiah Gould, Cornell University Press, Ithaca.

Alexander of Aphrodisias (1999b), *On Aristotle's **Prior Analytics** 1.14-22*, translated by Ian Mueller with Josiah Gould, Cornell University Press, Ithaca.

Alexander of Aphrodisias (2001), *On Aristotle's **Topics** 1*, translated by Johannes M. Van Ophuijsen, Cornell University Press, Ithaca.

Ammonius (1998), *On Aristotle's **On Interpretation** 9* with Boethius *On Aristotle's **On Interpretation** 9*, translated by David Blank and Norman Kretzmann, Cornell University Press, Ithaca.

Aristotle (1908), *Metaphysica*, translated by W.D. Ross in W.D. Ross (ed.), *The Works of Aristotle translated into English*, Vol.VIII, Oxford University Press, London.

Aristotle (1928), *Topica and De Sophisticis Elenchis*, translated by W.A. Pickard-Cambridge in W.D. Ross (ed.), *The Works of Aristotle translated into English*, Vol.I, Oxford University Press, London.

Aristotle (1963), *Categories and De Interpretatione*, translated with notes by John Ackrill, Clarendon, Oxford.

Aristotle (1975), *Posterior Analytics*, translated with notes by Jonathan Barnes, Clarendon, Oxford.

Aristotle (1989), *Prior Analytics*, translated by Robin Smith, Hackett, Indianapolis.

Averröes (1562), *Quaesita octo in lib. Priorusm*, in *Aristotelis Opera cum Averrois Commentariis*, Vol.I Part 2b u. 3 (Venetiis apud Junctas 1562-1574) reprinted Minerva, Frankfurt.

Avicenna (1984), *Remarks and Admonitions Part One: Logic*, translated from the original Arabic with an introduction and notes by Shams Constantine Inati [= *Mediaeval Sources in Translation*, Vol.28], The Pontifical Institute of Mediaeval Studies, Toronto.

Becker, A. (1933), *Die Aristotelische Theorie der Moglichkeitsschlüsse: eine logisch-philologische Untersuchung der Kapitel 13-22 von Aristoteles **Analytica priora I***, Junker und Dünnhaupt, Berlin.

Boethius, A.M. Severino Boezio (1969), *De hypotheticis syllogismis, testo, traduzione, introduzione e commento*, (ed.) Luca Obertello, Paideia, Brescia.

Boethius (1988), *In Ciceronis Topica*, translated with notes and an introduction by Eleanor Stump, Cornell University Press, Ithaca.

Boethius (1998), See Ammonius (1998).

Buridan, J. (1499), *Perutile compendium totius logice Joannis Buridani cum preclarissima solertissimi viri Joannis Dorp expositione*, Venice; reprinted Minerva, Frankfurt.

Buridan, J. (1985), *Jean Buridan's Logic: the treatise on supposition, the treatise on consequences*, translated with a philosophical introduction by Peter King, Reidel, Dordrecht.

Buridan, J. (2001), *Summulae de dialectica*, an annotated translation with a philosophical introduction by Gyula Klima, Yale University Press, New Haven.

Campsall, R. (1968), *Questiones super librum priorum analeticorum*, edited by Edward A. Synan, The Pontifical Institute of Medieval Studies, Toronto.

Elamrani-Jamal, A. (1995), 'Ibn Rušd et les *premiers analytiques* d'Aristote', *Arabic Sciences and Philosophy*, Vol.5, 51-74.

Garlandus Compotista (1959), *Dialectica*, first edition of the manuscripts with an introduction on the life and works of the author and on the contents of the present work by L.M. de Rijk, Van Gorcum, Assen.

Hughes, G.E. (1989), 'The modal logic of John Buridan', in G. Corsi, C. Mangione and M. Mugnai (eds.), *Atti del Convegno Internazionale di Storia della Logica. Le teorie della modalità*, CLUEB, Bologna, 93-111.

Hughes, G.E. and Cresswell, M.J. (1968), *An Introduction to Modal Logic*, Methuen, London.

Jacobi, K. (1980), *Die Modalbegriffe in den logischen Schriften des Wilhelm von Shyreswood und in anderen Kompendien des 12. und 13. Jahrhunderts: Funktionsbestimmung und Gebrauch in der logischen Analyse*, Brill, Leiden.

John of Salisbury (1962), *Metalogicon*, translated by Daniel D. McGarry, University of California Press, Berkeley.

Kilwardby, R. (1516), *Reverendi Magistri Egidii Romani in libros Priorum analecticorum Aristotelis Expositio*, Venice, reprinted Minerva, Frankfurt 1968.

Knuuttila, S. (1993), *Modalities in Medieval Philosophy*, Routledge, London.

Lagerlund, H. (2000), *Modal Syllogistics in the Middle Ages*, Brill, Leiden.

Lewis, C.I. and Langford, C.H. (1959), *Symbolic Logic*, Dover, New York.

Marenbon, J. (2000), *Aristotelian Logic, Platonism and the Context of Early Medieval Philosophy in the West*, Ashgate, Aldershot.

Nortmann, U. (1996), *Modale Syllogismen, mögliche Welten, Essentialismus: eine Analyse der aristotelischen Modallogik*, Walter de Gruyter, Berlin.

Ockham (1962), *Summa Logicae Pars Secunda et Tertiae Prima*, edited by Philotheus Boehner, The Franciscan Institute, St. Bonaventura NY.

Patterson, R (1995), *Aristotle's Modal Logic: essence and entailment in the **Organon***, Cambridge University Press, Cambridge.

Plato (1993), *Phaedo*, translated by David Gallop, Oxford University Press, Oxford.

Porphyry (1975), *Isagoge*, translation, introduction and notes by Edward W. Warren, The Pontifical Institute of Medieval Studies, Toronto.

Rescher, N. (1967), *Temporal modalities in Arabic logic* [= *Foundations of Language Supplementary series*, Vol.2], Reidel, Dordrecht.

Rini, A. (2000), '*Hupo* in the *Prior Analytics*: a note on Disamis XLL', *History and Philosophy of Logic*, Vol.21 pp.259-264.

Stephanus (2000), *On Aristotle's* ***On Interpretation***, translated by William Charlton with 'Philoponus', *On Aristotle's* ***On the Soul 3.9-13***, Cornell University Press, Ithaca.

Street, T. (2002), 'An outline of Avicenna's syllogistic', in *Archiv für Geschichte der Philosophie*, Vol.84 pp.129–160.

Street, T. (n.d.), Unpublished translation of Ibn Sīnā, *al-Isharat* with Tūsī's commentary.

Street, T. (2000), 'Avicenna and Tūsī on the contradiction and conversion of the absolute', in *History and Philosophy of Logic*, Vol.21 pp.45-56.

Thom, P. (1981), *The Syllogism*, Philosophia, München.

Thom, P. (1996), *The logic of essentialism: an interpretation of Aristotle's modal syllogistic* [= *New Synthese Historical Library*, Vol.43], Kluwer, Dordrecht.

Thom, P. (1997), 'Syllogismus / Syllogistik', in J. Ritter and K. Gründer (eds.), *Historisches Wörterbuch der Philosophie*, Vol.10 pp.687-707.

Thom, P. (1998), 'How to base apodeictic syllogistic on essentialist theory', in *Philosophiegeschichte und logische Analyse*, Vol.1. pp. 171-185.

Index

For Product Safety Concerns and Information please contact our EU representative GPSR@taylorandfrancis.com Taylor & Francis Verlag GmbH, Kaufingerstraße 24, 80331 München, Germany

Printed and bound by CPI Group (UK) Ltd, Croydon, CR0 4YY
14/09/2026
02216223-0017